MW00477127

COSMIC
COLLISIONS

DANA DESONIE

Foreword by David H. Levy
and Carolyn and Eugene Shoemaker

A SCIENTIFIC AMERICAN FOCUS BOOK

Henry Holt and Company
New York

Henry Holt and Company, Inc.
Publishers since 1866
115 West 18th Street
New York, New York 10011

Henry Holt® is a registered trademark
of Henry Holt and Company, Inc.

Library of Congress Cataloging-in-Publication Data
Desonie, Dana.
Cosmic collisions / by Dana Desonie; foreword by
David Levy and Carolyn & Eugene Shoemaker.—1st ed.
p.   cm.—(A Scientific American focus book)
Includes index.
1. Asteroids.   2. Catastrophes (Geology).
3. Cryptoexplosion structures.   I. Title.   II. Series.
QB377.D48   1995   95-2440
551.3'.97—dc20   CIP

ISBN 0-8050-3843-4
ISBN 0-8050-3844-2 (An Owl Book: pbk.)

Henry Holt books are available for special promotions
and premiums. For details contact: Director, Special Markets.

First Edition—1996

Conceived by Robert Ubell Associates, Inc.
Project Director: Robert N. Ubell
Project Manager: Luis A. Gonzalez
Art Direction: J.C. Suarès
Design: Amy Gonzalez
Production: ChristyTrotter

Printed in the United States of America
All first editions are printed on acid-free paper.∞

10   9   8   7   6   5   4   3   2   1
10   9   8   7   6   5   4   3   2   1   (pbk.)

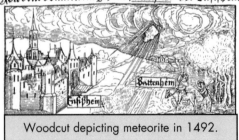

Woodcut depicting meteorite in 1492.

C O N T E N T S

6
Foreword
David H. Levy and Carolyn
and Eugene Shoemaker

10
Chapter One
Earth's History of
Cosmic Collisions

14
Chapter Two
Earth's Infancy

24
Chapter Three
Early Earth and Moon

34
Chapter Four
Identified Flying Objects:
Asteroids and Comets

46
Chapter Five
Meteorites—They Strike
Planets, Don't They?

56
Chapter Six
Earth's Ancient Wounds

66
Chapter Seven
Demise of the Dinosaurs

80
Chapter Eight
Out with the Old World,
In with the New

88
Chapter Nine
Twentieth-Century Hits

96
Chapter Ten
Future Shock

106
Chapter Eleven
Identifying
Flying Objects

114
Chapter Twelve
Defensive Maneuvers

120
Chapter Thirteen
The End?

124
Photo Credits

126
Index and
Further Reading

128
Acknowledgments

By David H. Levy and Carolyn & Eugene Shoemaker

n July 16, 1994, the first fragment of Comet Shoemaker-Levy 9 entered Jupiter's stratosphere traveling at the speed of 60 kilometers per second— that's about 134,000 miles per hour. Seconds later it exploded. A plume of shock-heated cometary material mixed with hot gases from Jupiter's atmosphere soared more than two thousand miles above the tops of Jupiter's clouds. As material from the plume fell back into the stratosphere of Jupiter, it left a large, crescent-shaped dark spot partly surrounding a smaller intense dark cloud that formed along the entry path of the comet.

Two days later, Fragment G hit Jupiter with such force that its plume was, in some wavelengths of observation, fifty times brighter than the entire planet. By the end of impact week, Jupiter was bruised with the markings of 21 impacts, dark scars larger than Earth. Not in the 400-year history of the telescope has any planet displayed such dramatic new features. Nine months after the impacts ended, Jupiter's southern hemisphere was graced with a belt of dark material, still easily visible through small telescopes.

We are only now beginning to grasp the magnitude of what has happened. What we witnessed was one of the solar system's basic events. It is the kind of event that, when repeated millions of times, led to the accumulation of the planets from smaller bodies when the solar system was young. It was the kind of event that, in the Earth's early years, may have deposited the carbon, hydrogen, oxygen, and nitrogen that ultimately led to the evolution of life on this planet, and that led to several mass extinctions, including that of the dinosaurs. What do cosmic collisions portend for Earth? Do we really have anything to worry about?

Had Comet Shoemaker-Levy 9 been as it was before it

broke apart, and had it hit Earth, our fate might have been similar to that of the dinosaurs.

There is a great deal of evidence, both in a thin world-wide iridium-rich layer and around the impact crater, to corroborate the following scenario: A comet (or asteroid) hit Earth just off the coast of modern-day Yucatan in eastern Mexico. In less than a minute the force of the impact carved out a crater some 100 kilometers wide and 30 kilometers deep, sending material high into the atmosphere. The sky glowed red and any creature in view of the sky would have felt temperatures as high as an oven set to broil. Material from the crater rained down everywhere. The red sky and searing temperatures lasted for more than two hours. Ground fires erupted all over the planet. Meanwhile, fine particles of dust settled in the stratosphere, darkening the sky so thoroughly that Earth's surface was as black as a photographic darkroom. Rain, rich with nitric and sulfuric acids, drenched the world. If the situation on Jupiter is any indication, darkness probably lasted more than a year, and a runaway greenhouse effect persisted for centuries afterward. More than seventy percent of all species of life perished.

Happily, we know of only one asteroid large enough to do that kind of damage to Earth—Comet Swift-Tuttle, whose trail of debris causes the Perseid meteor shower every August, is large enough, and it can cross the orbit of Earth. According to astronomer Brian Marsden, Swift-Tuttle is the largest known comet with the potential to strike the Earth sometime in the next several thousand years. Although the chances for it to hit are almost infinitesimal, it is prudent to keep a close eye on its path as it cruises through the solar system.

What about asteroids or comets we have not discovered? There is always a chance that some intruder could hit us with virtually no warning. There are more than 2000 asteroids (a kilometer or more in diameter) that are in orbits that could someday lead to collisions with Earth, and after a decade of serious searching, we know of about a twentieth of these.

A coordinated program to find these objects could begin in earnest within the next few years. Within a quarter century, we should know if any such intruders exist. As for the comets, with their long looping orbits that take them to the edge of the planetary system and back, we would always have to keep watch. While the history of cosmic collisions is as old as our solar system itself, only recently have comets received the attention they deserve. It is ironic that the alarm came so recently, for a look at the Moon with the smallest telescope will reveal a surface pockmarked with the remains of ancient impacts. The Moon keeps its crater record, but we know that Earth, a much larger target, has been hit more often. We are now alert to the role of impacts. As we become more aware of how serious the impact problem is, the coming years should be exhilarating.

Earth's History of
Cosmic Collisions

t is the year 2110. Kali, a giant peanut-shaped asteroid named for the Hindu goddess of destruction, hurtles towards Earth. At stake is human civilization, and possibly human life itself. A brave crew of astronauts, in a Mars-based starship, intercepts Kali. On her surface, they place a mass driver, a device that over the next several months will mine enough material from the asteroid to change her mass and alter her orbit just enough to bypass Earth. They hope.

This story is from Arthur C. Clark's novel *The Hammer of God*, but it is not just his fantasy. There is some truth in all the science fiction novels, *Star Trek* episodes, and even *The Simpsons* cartoon shows in which a planet is threatened by a cosmic collision. Springfield really is in danger, as is Tokyo, Casablanca, Rio, Auckland, Baghdad. Civilization could fall to an asteroid whose name we do not yet know.

Throughout time, cosmic collisions have sculpted our planet and its life. In fact, were it not for asteroid impacts, there would be no heavenly bodies—no Earth, no Moon. The Earth itself is composed of countless asteroids fused into a single sphere. The Moon is Earth's child, thrown from her side when a giant meteorite struck. Even the gases necessary for life may have been delivered by collisions with comets.

Collisions have molded the surfaces of the inner planets. Although the Earth's smooth face little resembles the Moon's pockmarked one, impact craters are found all over the world, from the relatively young Meteor Crater in Arizona, to the two-billion-year-old Sudbury Structure in Canada. Debris from collisions litters Earth's surface, from the glassy spheres that result from impacts, to the 54-metric-ton iron meteorite, Hoba, embedded in a crater in southwestern Africa.

Were it not for impacts, life on Earth would be very different—dinosaurs might still rule the world and humans might not even have made it to Nature's drawing board. One day, more than 65 million years ago, the collision of an enormous meteorite with Earth likely triggered the extinction of the dinosaurs, as well as two-thirds of the Earth's other species. Loss of so many species wiped the biological slate almost clean, paving the way for mammals to rise to dominance and, later, for the ascendancy of one notable mammalian species, human beings. A few scientists think these mass extinctions are cyclic, that Earth has repeatedly met its Nemesis—and will again.

Extinction of the dinosaurs—one minute following meteorite impact with Earth circa 65 million years ago.

Meteorites continue to shell the Earth. In this century, a woman in Alabama, a forest in Siberia, and a car in New York, to name a few, all had run-ins with these cosmic invaders. Although nothing very large has struck Earth lately, we know that massive objects do collide with planets. We had a front-row seat in July 1994 when more than 20 fragments of Comet Shoemaker-Levy 9, each about one kilometer in diameter, bombarded the planet Jupiter.

The orbits of many asteroids and comets bring them near the Earth. Most of these visitors from the cosmos fall into the Sun or are flung from the solar system by the gravitational field of a massive planet. But for each meteorite or comet that vanishes or strikes a planet, there is another waiting deep in space, and each is a potential threat to Earth.

We are just beginning to survey the sky for these threats—locating them, calculating their orbits, wondering if one might come just a little too close. We are also starting to talk about ways to deal with a potential collision. Will we know it's coming? How long in advance? Should we evacuate the targeted region, deflect the meteor, or attempt to blow it up? Will we have the time to destroy it? Will we have the capability?

overleaf:
Aerial view of
Meteor Crater,
Arizona.

Earth's Infancy

 ur universe has never been static. From the moment of its formation, matter was in motion; it darted and flew and, ultimately, collided with other matter. The Sun and its planets, including the Earth itself, are assortments of cosmic stuff that crashed together to create the celestial bodies, in a process called accretion. From the very beginning, cosmic collisions have been essential to creation.

Origin of the Universe

The universe came to be in an explosion scientists call the "Big Bang." In a single moment, perhaps 15 billion years ago, all the matter and energy of the cosmos were concentrated into a space smaller than a dime. From that tiny and infinitely hot dense speck, the universe expanded and began to cool.

The first distinctive materials to appear in the new universe were tiny subatomic particles, the fundamental building-blocks of all sub-

Periodic table of the elements.

The accretion of planetesimals into the planets and Sun.

stances. These particles soon came together to make atoms of the two lightest elements, hydrogen and helium. Although the universe was (and is) still expanding, hydrogen and helium gases gathered as giant clouds, which eventually became the galaxies. The first stars were created within these galaxies.

The air we breathe, the water we drink, the stones we stand on, and the metal we shape into tools—all the essential components of our world—began as that hydrogen and helium. Hydrogen and helium were forged into an enormous variety of chemical elements in the superdense, superhot cores of dying stars. The existence of the other elements is a clue to scientists, indicating that many stars were born and died before the genesis of our Sun.

A Solar System Is Born

Most stars shine because of the nuclear energy released as they convert hydrogen to helium. Just like a car's, a star's engine has limited fuel, and it eventually depletes its hydrogen supply. Roughly five billion years ago, near our solar nebula—the cloud of matter that was to become our solar system—a star many times more massive than the Sun ran out of hydrogen in its core. Without energy to hold itself together, the star caved in. Curiously, that collapse generated another

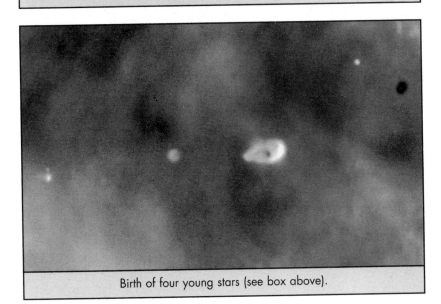

Star cradle in the Great Nebula in Orion through the
Hubble Space Telescope.

Birth of four young stars (see box above).

type of energy, called *gravitational potential energy*, which is the energy created by an object moving under the force of gravity. The liberation of enormous amounts of gravitational potential energy caused the star to become so hot that its helium ignited. In that inferno, all the middle-weight elements, from lithium through iron, were created.

Eventually, the star's helium stores were also consumed, and its core experienced a violent collapse, sending a shock wave throughout this region of the Milky Way galaxy. The wave's energy swiftly forged the heavier elements, such as gold and uranium. Heat generated by the nuclear reactions that converted the light and medium elements into the heavy ones set off an explosion like a galactic atomic bomb—a supernova. The explosion burst apart the star, shooting its newly created elements into the nebula beyond. The tremendous shock wave that accompanied the supernova explosion fragmented and compressed the nebula into giant clouds.

In one nearby cloud, the one that was to become our solar system, passage of the shock wave caused incredible turbulence, which resulted in the liberation of many types of energy. Gravity pulled the moving matter inward, releasing gravitational potential energy. As the particles of dust and gas came closer together, small particles collided to form larger ones, releasing kinetic energy. A body in motion possesses kinetic energy, which is freed when it stops. Kinetic energy is very important in the story of cosmic collisions.

Several processes fed on each other, leading to the formation of our solar system. Gravity continued to attract matter towards the cloud's core, which in turn further increased the core's gravity. Angular momentum, rotational energy of a spinning object, caused the cloud to slowly spin. Although most of the cloud's mass migrated to its center, the remaining matter formed a great disk that spiraled outward. In this disk were all the chemical elements of the solar system—the initial hydrogen and helium from the Big Bang, plus the elements created in many supernova explosions. Even simple organic molecules, the stuff of which life is made, were present in the cloud.

As the solar disk continued to collapse, new bursts of released energy kept the temperature climbing. After about 50 million years, the core of the disk became so hot that nuclear reactions, the conversion of hydrogen to helium, began. The Sun ignited.

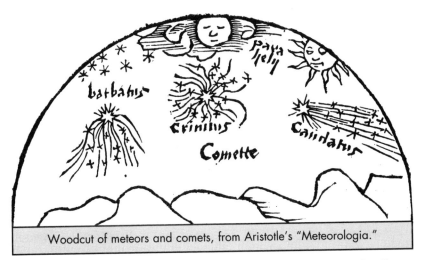

Woodcut of meteors and comets, from Aristotle's "Meteorologia."

As gas pressure built up in the burning star, gravitational collapse ceased. Temperature differences across the solar disk were extreme, ranging from more than 20°C in the center to -270°C at the outer reaches, the temperature of the outer solar system today. Matter condensed in the cooling disk, with gravity attracting the heaviest elements to its center. These metallic and rocky compounds became the constituents of the inner planets. At the same time, light gaseous elements, called volatiles, that existed near the Sun, vaporized. Volatile elements were stable in the colder outer portions of the disk where they bound together into compounds such as water, methane and ammonia.

In the solar disk, molecules collided and fused into specks, and these specks in turn crashed and coalesced into pebbles, rocks and boulders. As the sizes of the projectiles grew, so did the intensity of their collisions. Within ten million years, boulders became planetesimals, rocky objects dozens of kilometers across. Larger planetesimals developed gravitational fields strong enough to attract smaller ones. Eventually only large bodies remained, and their collisions ultimately created the planets. The inner planets grew, or accreted, by the addition of metal and rock bodies in the inner disk. The outer planets formed by agglomeration of planetesimals of rock and ice further out in the solar system.

Our Sun and its planets were constructed simultaneously about 4.6

billion years ago. Scientists think planetary accretion took about 100 million years, more for the larger outer planets and less for the smaller inner planets.

The Planetary Bodies

For the Earth to grow from cosmic dust to its present size in roughly 70 million years, the impact rate must have been one billion times what it is today. Enormous impacts were commonplace, with a violent collision occurring at least once a month. Rock and dust thrown up during collisions blocked the light of the newly formed Sun for months at a time. This dark, waterless, lifeless and exceedingly hot planet bore little resemblance to our Earth today.

Artist's concept of the solar system.

In nearby space, the situation was similar. At regular distances outward from the Sun, planets formed. The inner planets, Mercury, Venus,

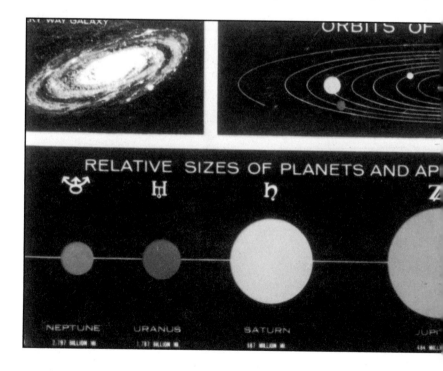

RELATIVE SIZES OF PLANETS AND AP

NEPTUNE URANUS SATURN JUP

Earth and Mars, were created of metal and rock materials in the forms of sulfides and silicates, respectively. The outer planets, Jupiter, Saturn, Uranus, Neptune and Pluto, were made primarily of hydrogen found in the compounds methane, ammonia, hydrogen sulfide and water. Each of the planets was the "winner" of innumerable encounters with lesser bodies, or maybe we can say that the colliders joined forces to form the planets.

The Sun and planets are the major bodies of our solar system, but many other types of objects reside here. The two groups that are most important to the cosmic-collision story are asteroids and comets. Asteroids are generally composed of the same stuff as the inner planets, primarily rock and metal. Most of them inhabit a gap in the otherwise regular spacing of planets; the gap lies between the orbits of Mars and Jupiter, and is known as the *asteroid belt*. Scientists believe that these asteroids are a failed planet, a planet that never formed because of grav-

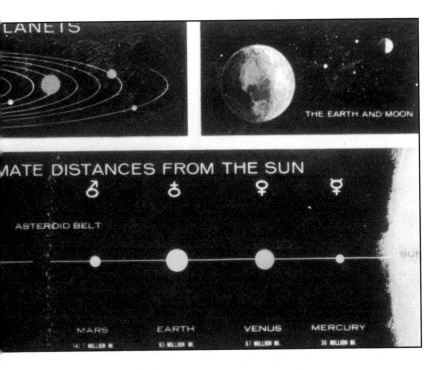

LANETS

THE EARTH AND MOON

MATE DISTANCES FROM THE SUN

ASTEROID BELT

MARS	EARTH	VENUS	MERCURY
141.7 MILLION MI.	93 MILLION MI.	67 MILLION MI.	36 MILLION MI.

SUN

itational disruption by Jupiter.

Comets live in the far outer solar system beyond Neptune. We know of their existence because some have highly elliptical orbits and pass through the inner solar system. Like the outer planets, they are made of rock and frozen water, methane, ammonia and hydrogen sulfide.

Differentiation of Earth and Other Planets

During its formation, the proto-Earth incorporated all the objects it encountered in its orbital path. But Earth today is not a collection of fragments—a rocky piece here, an icy part there, an asteroid here and a comet there. The processes of homogenization and differentiation layered the planet into its present core, mantle and crust structure.

Planetary accretion resulted in the widespread availability of heat energy to the early planets. Kinetic energy was released by the collisions of dust, rocks, boulders, asteroids and comets. Gravitational potential energy was liberated by the additional mass of the accreted bodies as they compressed towards the planet's center. Many elements in the early solar system were radioactive, and nuclear energy was emitted with each radioactive decay. All of this energy turned into heat. The planetary cauldrons grew so hot that each of the inner planets melted. A homogeneous soup made of the liquefied stony minerals, metals and sulfides from all the planetesimals, asteroids and comets came together to make each planet.

After homogenization, the planets began to differentiate. Just as a coin sinks to the bottom of a pool of water, gravity pulled the heaviest molecules towards the planet's center. These dense, metal-rich liquids, primarily iron and nickel, became the Earth's core. The siderophile, or iron-loving, elements followed. One very important siderophile element, iridium, so completely entered the core that when it was found in abundance in a 65-million-year-old rock layer on the surface, it sparked a new controversial theory about dinosaur extinctions.

Out of the remaining soup, the lighter stony liquids rose and solidified to become the Earth's crust. Heavier stony liquids remained in the Earth's interior and became the mantle. Volatile compounds, such as water and carbon monoxide, floated to the Earth's surface but were vaporized and lost. This differentiation of the planet into distinct layers of crust, mantle and core created the structure of the Earth, the other inner planets and the large asteroids.

The layered structure of the Earth is essential to life as we know it. If the Earth had no core it would have no magnetic field. Imagine the complications in navigation! Differentiation is most important to our understanding of meteorites. Many meteorites are representative of the core, mantle or crust of the asteroids they came from.

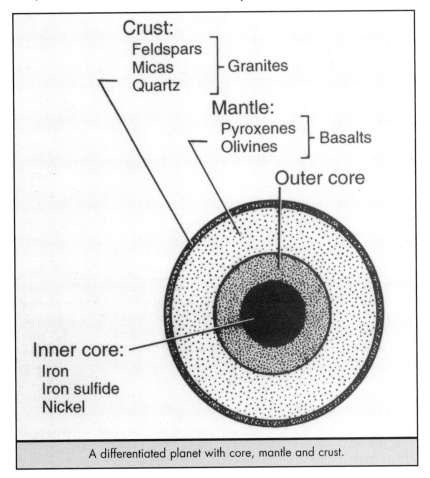

A differentiated planet with core, mantle and crust.

Early
Earth and Moon

he Moon is a most unusual celestial object. Of the inner planets, probably only Earth and Mars have satellites. Those of Mars are probably captured asteroids measuring up to only 20 kilometers across. But our Moon is more than one-quarter the diameter of the Earth and scientists believe it was born in the most violent cosmic collision our planet has ever seen. We can thank this collision for the intensity of Earth's tides and for lighting our most romantic nights. The Moon also keeps the best records we have of our own planet's early history.

Birth of the Moon

A few tens of million years after the planets accreted, things in the Earth's region of space were settling down. Fewer objects littered the planet's orbital path and collision rates decreased. Differentiation had separated the Earth into core, mantle and crust. The planet may have developed the beginnings of an atmosphere. But where was the Moon?

The question of the Moon's origin has been difficult for scientists to answer. Several years after the Apollo missions brought rock and soil samples from the lunar surface to Earth's laboratories, a satisfactory history had not yet been agreed upon. Did the Moon grow alongside its sister Earth? Did Earth simply capture a passing asteroid as had Mars? At some time, did the Earth spin so fast that it flung a piece of itself into space? A winning theory had to account for several lines of evidence indicating that the Earth and Moon were once one. The composition of the Moon is very much like that of the Earth's mantle, and for a body its size and distance from the Sun, the Moon has very little core. Oxygen isotope ratios of the Earth and Moon are the same, meaning that they originated in the same part of the solar system. What's more, the Moon lacks gaseous volatile elements. If volatiles were present in the lunar interior, as they are in the Earth's, the Moon could have a small atmos-

Painting depicting probable impact of giant projectile with Earth resulting in the formation of the Moon.

phere, which it doesn't. Also intriguing to scientists was Earth's abnormally fast axial spin, far faster than would be expected for a planet its size and distance from the Sun.

Photo of craters on the far side of the Moon taken by Apollo 10.

Among the lunar scientists, consensus has grown up around an idea first seriously discussed in 1984—the *impactor* theory. Recent computer models of the scenario make the theory seem more likely—and lively. About 4.5 billion years ago, a giant planetesimal, probably the size of Mars, struck the Earth in a monumental cosmic collision. The crusts of both bodies were thin and offered little resistance to the crash. The impactor plowed through the Earth's mantle, all the way to the core, where much of it remains. Silica-rich rocks from the mantles of both bodies vaporized and joined additional debris in a cloud encircling the Earth. As the gases cooled, they condensed into a thin ring. Here again, collisions caused the particles to aggregate, and within a few tens of millions of years, our Moon was born.

So much energy was released during the Moon's rapid accretion that, as the lunar sphere grew, it melted to a depth of a few hundred kilometers, forming a magma ocean. Fiery lava tides were created by the nearby Earth's gravitational pull. Many geologists believe that the

Painting of Earth and Moon shortly after Moon's birth.

Earth had a similar molten lava sea during accretionary times, but no geologic record of it remains.

The impactor theory accounts for all important observations. If an impactor hit the Earth off-center, it would have increased the planet's orbital speed. That the Moon is atmosphere-free is consistent with the vaporization of volatiles after the collision. Although the impactor had a core, it remains near the Earth's core, and the Moon is less dense than other planetary bodies its size.

Consistent with the impactor theory is the fact that the Moon is moving away from Earth, at about three centimeters per year. Initially the ring of impact debris was only a few Earth diameters away. With time the satellite was moved outward by tidal forces. Imagine a moonrise 4.5 billion years ago, when the Moon appeared more than forty times larger than it does today.

The Era of Intense Bombardment

The intensely cratered lunar surface could hardly be more different from the face of its host planet. The near side of the Moon is home to more than 30,000 craters, while the entire Earth has fewer than 200. Studies of the lunar surface have shown that the cratering rate has been roughly constant for at least 3.2 billion years. Between 3.2 and 3.85 billion years ago, bombardments were approximately a thousand times more common than they are now. More than 3.85 billion years ago, the rate was four thousand times greater than today. The intensity of collisions undoubtedly had an effect on the early evolution of continents, oceans, atmosphere and life.

By studying the geologic history recorded on the Moon's surface, scientists can tell a great deal about the Earth's early existence. The abundance of craters on the lunar landscape suggests profound bombardment in this region of space. Since the Earth is more massive and has greater gravity than the Moon, it has pulled many more impactors towards it. Many of the impactors now break up in the Earth's atmosphere but, in the early years, Earth's impact rate must have been greater than the Moon's. It is our planet's plate-tectonic activity that has destroyed its bombardment record.

Our active Earth—intense folding in the Andes mountains of northwestern Argentina.

Destroyer of Early Earth History

To understand the very early history of our world, earth scientists had to travel (figuratively speaking) to the Moon. Why on Earth—a planet full of exciting geological phenomena, like volcanoes, earthquakes, mountain-building, and erosion (the wearing-away of the surface of the Earth)—did they need rock samples from our satellite to piece together the story of our planet's beginnings? It was Earth's interesting geological processes that made lunar expeditions necessary. Geologic records of impact events have been all but obliterated on Earth.

The Earth is four times the diameter of the Moon. A larger planetary body acquires more heat and retains it more effectively. Earth's abundant heat comes from having six times more surface gravitational energy and from taking on the kinetic energy of many more planetesimals. Earth also contains many more radioactive elements, each emitting energy as it decays. Internal heat acts as an engine that drives the planet's geological activity.

The theory of plate tectonics describes the movements of large slabs of the Earth's crust around its surface. Just as a pot of soup being heated from below develops convection cells (warmer soup rises and cooler soup sinks), differences in temperature result in convection cells in the planet's mantle. Convection drives the Earth's crustal plates around —sometimes they separate, sometimes they collide, and sometimes they just move past one another. These encounters can create volcanoes to form new young rocks. Plate movements can also destroy or change old rocks and features, or they can smash sections of crust together to form mountain ranges. A great deal of the geological activity on Earth can be attributed to plate tectonics.

Earth's rocks and features are continually being destroyed and recreated in a giant planetary recycling program. The oldest rocks yet discovered on Earth are about four billion years old, and rocks older than 3.5 billion years are very rare. The ages of most geologic terrains, including all the ocean basins, are counted in mere millions of years.

By contrast, the Moon's heat engine fizzled out in its infancy; lunar rocks and structures are not recycled. That means virtually all lunar samples are old; many are more ancient than four billion years. Only impacts continue to affect the geology of the lunar surface, creating craters and melting rocks that cool into rock formations called *impact basalts*. Fortunately for us, our geologically frozen Moon has a lot to tell scientists about the early Earth.

Annihilation of Earth's Early Atmosphere, Oceans and Life

During Earth's fiery infancy, gases vaporized and were lost. But after some planetary cooling, a primitive gaseous atmosphere could have been retained. Where did these gases come from? Erupting volcanoes released volatile elements once trapped in the Earth's interior; these gases made up most of the early atmosphere. However, volcanic gases are rich in sulfur and other toxic chemicals, so they could not have been the sole sources of atmospheric gases. If they had been, Earth's environment would have been hostile to life.

Some atmospheric gases were delivered from outer space. When comets from the outer reaches of the solar system collided with Earth, they donated gases to Earth's immature atmosphere. The frozen gases found in comets, primarily water and carbon dioxide, are necessary for life as it exists on our planet. Scientific missions to Halley's comet revealed that its hydrogen-isotope ratio is the same as that of terrestrial ocean waters, evidence that comets did indeed supply some of Earth's volatiles. Once a preliminary atmosphere was created, some gases would have condensed to form rain, beginning the early oceans. Also abundant in comets are complex organic compounds, which some scientists have proposed were the seeds of life on Earth.

The atmosphere and oceans, even life itself, may have evolved before the end of the era of intense bombardment. A collision with a large enough object would have dissipated the Earth's atmosphere and possibly boiled away its ocean. Smaller collisions could have partially destroyed the atmosphere and evaporated the upper ocean layers. Had life evolved in this unstable environment— and scientists believe that it could have, many times—it would have been destroyed as Earth was sterilized by impact events. Eventually, however, the bombardment rate decreased, the atmosphere and oceans stabilized, and a complex system of life arose.

overleaf:
Earth activity—fire fountain of Pu' volcano, Hawaii.

Identified Flying Objects:
Asteroids and Comets

In the grand era of cosmic collisions, early in the life of the solar system, rocks pummeled boulders that struck planetesimals that eventually accreted into the planets. Even after the planets formed, bodies large enough to cause catastrophic collisions roamed the skies. But now, billions of years later, the planets have swept their orbits clean of rocky debris. No objects threaten to bombard the Earth today. Or do they? What objects pose a danger to Earth and the other planets? Do potential colliders orbit near the Earth? Is there a warehouse for these impactors somewhere in the solar system?

There are now two types of threatening objects: asteroids and comets, collectively called *small bodies*. The asteroid belt is jammed with bodies that can be perturbed into a near-Earth orbit by nearby Jupiter, while comets prowl the outer solar system. Any object that strikes the Earth is called a *meteorite*. What are the sources of meteorites and where do they come from?

A sample of the crust of Asteriod Vesta—one of only three extraterrestrial bodies from which we have samples.

The Missing Planet

Asteroids are too small and distant to be seen with the unaided eye and so were still unknown long after ancient observers first turned their eyes heavenwards. The planets were discovered because they wandered back and

forth against the background of fixed stars. Comets, although rare, were quite conspicuous and were regarded with awe and terror because of their supernatural appearance. But asteroids were only discovered in the eighteenth century and, at first, only mathematically.

In 1766, the German astronomer Johann D. Titus noticed a gap in the otherwise regular spacing of the planets, and suggested that a body should exist between Mars and Jupiter, about 2.8 astronomical units from the Sun. One astronomical unit, or AU, is the mean distance of the Earth from the Sun (about 149.6 million kilometers, or 93 million miles). Titus initiated the search for the missing planet, which culminated in the discovery of a planetary object by Guiseppe Piazza in 1801. But the object, named Ceres, was too small to be a planet. Within a few years, many more planetesimals were discovered in the same region of space and were collectively named *asteroids*. Since then, astronomers have identified about 18,000 of these small bodies.

Asteroids are cold and lifeless. They may be spherical or elongate but are usually irregularly shaped. A few may be double or multiple bodies, and some orbit each other. Their surfaces are altered by the meteorites, dust, solar wind and cosmic rays that continually bombard them. They travel through the cosmos at a typical speed of about 20 kilometers per second.

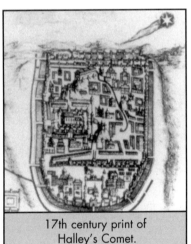

17th century print of Halley's Comet.

The Asteroid Belt

The broad zone between 1.8 and 4.3 AU from the Sun, named the *asteroid belt*, contains innumerable planetesimals with a total combined mass less than that of our Moon's. Astronomers calculate that the mass of the missing planet should be 2.8 times the mass of the Earth. They speculate that in the long-distant past, Jupiter's gravity flung asteroids equal to two Earth-sized planets out of the solar system.

Asteroids of the main belt vary in size and relative abundance. The three largest account for about one-half the mass of the belt. Ceres, with a diameter of 930 kilometers, contains about a quarter of the mass of all known asteroids. More than 200 of the bodies are larger than 100 kilometers across and roughly a thousand are larger than 30 kilometers. There are an estimated one million with diameters of a kilometer or more. Any one of these could do considerable harm if it were to strike a planet.

Photo of Asteroids Ida and Dactyl taken from *Galileo* spacecraft.

Scientists blame the gravitational influence of Jupiter for the failure of main-belt asteroids to form a planet. Although these asteroids were probably very similar to the ones that accreted into the inner planets, the gravitational pull exerted by giant Jupiter forced them into high-velocity elliptical orbits. Instead of growing larger by accretion, the speed with which the asteroids collided caused them to break into bits. Main-belt asteroids often live in families, with several members sharing similar orbits and composition. These families are probably the offspring of a larger parent asteroid that fragmented in a crash.

Portion of the Bayeaux Tapestry, 1066 A.D.

Collisions in the asteroid belt are now infrequent, but no less violent. Small impacts generate craters and cracks. Encounters between larger asteroids usually result in fragmentation of both. Resulting bits can be tossed from their orbits or may remain in main-belt families.

Most asteroids dwell in stable, near-circular orbits within the asteroid belt. Sometimes Jupiter's gravity pulls an asteroid into such an elliptical orbit that it voyages through the inner solar system and is transformed into a near-Earth asteroid. If a near-Earth asteroid's new orbit crosses our planet's path, a collision may result.

Asteroid Composition

Asteroids were formed by the same processes that made the inner planets, and so they provide scientists with important information about planetary structure and development. In fact, some of what we know about planetary differentiation has been learned through observations of asteroids and the meteorites that come from them. For example, although scientists cannot directly observe the Earth's core, they can study asteroids that came from the cores of their own parent bodies.

Although it may seem odd that objects as inaccessible as asteroids are classified based on their composition, clues such as the body's density, the color of light reflected off its surface, and the chemistry of meteorites that were once asteroids, have made possible just such a system. Asteroids are grouped into three superclasses—igneous, metamorphic, and primitive. Each superclass is located at a different distance from the Sun within the asteroid belt. Regional variations in composition are the result of the extreme temperature gradient that was present in the ancient solar system.

Asteroids located near the inner part of the asteroid belt are made of the metallic and rocky materials that condensed at high temperatures. Igneous asteroids are composed of minerals similar to those that cool from molten lava, indicating that they were once completely liquid. Many are large enough to have undergone differentiation, the separation into a metallic core, rocky mantle, and crust. Asteroids found in the belt's outer reaches are termed "primitive" because they contain material unaltered since the beginning of the solar system, including volatiles and organic matter. Their surfaces are coated with carbon-rich minerals, giving them a black appearance. Although metamorphic asteroids were never hot enough to melt, they are depleted in volatiles and have little water, indicating some exposure to high temperatures. They reside in the central region of the belt.

Asteroid Ida

On its journey through the solar system, the U.S. spacecraft *Galileo* has been taking family portraits—asteroid family portraits, that is. In our first ever view of an asteroid, *Galileo* initially met and photographed 951 Gaspra, a solitary asteroid of the main belt 23 kilometers in diameter. Its second rendezvous was with 243 Ida, a larger, irregularly shaped (56 by 24 by 21 kilometers) body found with a tiny companion, Dactyl. The satellite is 1.4 kilometers in diameter; its color, reflectivity and texture indicate that it is Ida's close relative. Initially, astronomers thought that Dactyl blasted into orbit when Ida collided with a smaller asteroid or, more likely, that both Ida and Dactyl were remnants of a collision involving a larger ancestral asteroid. Closer photos reveal that Dactyl is quite spherical, breaking the solar system's rule that small bodies are irregularly shaped. To account for the sphericity, scientists now consider Dactyl to be a loose aggregate of boulders, gravel and sand agglomerated by its own feeble gravity. *Galileo* is now headed for an orbit around Jupiter, where it will drop probes into the planet's atmosphere and take photos for the rest of its useful life.

Comets: Faster than a Speeding Bullet

Unlike the lusterless asteroids, comets are a most spectacular celestial phenomenon—revered and feared in writings, tapestries, songs and paintings for more than two millennia. Their bright tails, which routinely stretch across space for several tens of millions of kilometers, have been either a source of beauty or an omen of bad tidings for much of human history. The earliest known citing of Halley's Comet appears in Chinese records from 240 B.C.

In spite of their almost supernatural status, comets are actually just icy asteroids. Scientists sometimes call them "dirty snowballs" since their nuclei are composed of grains of rock bound together by ices, including frozen water, carbon dioxide and carbon monoxide. Scientists once thought that an average comet's nucleus had a total diameter of only one to three kilometers, but photographs of the nucleus of Halley's Comet revealed a much larger, darker and more irregularly shaped sur-

Nucleus of Comet Halley photographed by Halley Multicolor Camera aboard *Giotto* spacecraft.

face than they had suspected. Now they think most comet nuclei are between five and ten kilometers in diameter.

Comets are noteworthy for their brightly lit heads and tails. As a comet races towards the Sun, the ice in its nucleus warms. Sublimation, the process that converts a solid into a gas without passing through a liquid phase, transforms the ice directly into mist. Heating cracks the nucleus, expelling frozen dust, with jets of sublimating gases to brilliantly reflect the Sun's light. Scientists think that comets can sublimate for about 500 journeys near the Sun before they run out of volatiles. A comet without volatiles is just an asteroid.

Most comets never experience sublimation. They are preserved in the deep-freeze of outer space, beyond the orbit of Neptune, and are barely affected by our far-distant Sun. These far-out comets have undergone little alteration, and contain some of the most primitive matter of the solar system.

Comet Warehouses: The Oort Cloud and Kuiper Belt

Each year, several previously unrecorded comets arrive in the inner solar system. Where do they come from? In 1950, Dutch astronomer Jan Oort concluded from his statistical study of comet orbits that although they originate far from the Sun, they are nevertheless a part of our solar system. He hypothesized that there was a reservoir of comets at a mean distance of about 20,000 AU, more than 500 times the distance from the Sun as the outermost planet, Pluto. Although this cometary warehouse, now called the *Oort cloud*, has never been seen, most astronomers think the evidence strongly favors its existence. As many as one hundred million to one trillion comets soar through the

Oort cloud. Because the Sun's gravitational influence is minimal at that distance, Oort-cloud residents are strongly influenced by passing stars, which may propel them far out into the Milky Way galaxy or may force them inwards toward the Sun.

At about the same time that Oort proposed the existence of the Oort cloud, American astronomer Gerard Kuiper proposed that another belt of comets, up to 10 trillion of them, orbited much closer to the Sun in deep space beyond Neptune. The first two members of the *Kuiper belt* were identified in 1992 and 1993 and several more have been added. The icy planetesimals of the belt did not accrete into a planet because of their low density and long orbital periods. Kuiper-belt comets can also be perturbed from their distant orbits and head toward the inner solar system. Beware—comets of the Kuiper belt can be huge, between 100 and 250 kilometers in diameter.

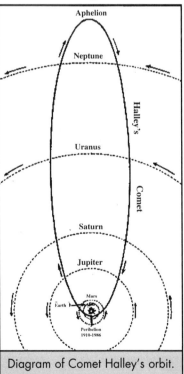

Diagram of Comet Halley's orbit.

Virtually all comets that venture towards the center of the solar system are burned up or are captured by the large planets Jupiter and Saturn. The few that get past those sentries develop orbits in the inner solar system and may someday collide with one of the inner planets. Some researchers think that comets travel in swarms, perturbed from their orbits en masse by the same passing star.

There are two varieties of comets that enter the inner solar system, and both are good candidates for cosmic collisions. About 600 comets have highly elliptical orbits around the Sun, taking millions of years to complete one trip. These so-called long-period comets spend most of their lives in the outer solar system, relatively unaffected by gravitational interactions with the planets. Because of their highly elliptical orbits, long-period comets travel rapidly through the inner solar system, between 50 and 60 kilometers per second. If a long-period comet happens to intersect Earth's orbit, it hits the atmosphere at very high velocity and stands a good chance of being obliterated before reaching the surface. If the comet makes it deep enough into the atmosphere, it can explode as an air blast and cause a tremendous amount of damage. It is thought that between 10 and 25 percent of the massive objects that threaten Earth are long-period comets.

Long-period comets can pose a danger to Earth in another way. They can be forced by a giant planet's gravity into a more circular orbit in the inner solar system. In other words, they can become short-period comets. Short-period comets can also originate in the Kuiper belt. The approximately 150 comets with orbits of less than 200 years spend most of their time in the inner solar system. Short-period comets move more slowly, between 30 and 40 kilometers per second in the vicinity of Earth's orbit, and are more likely to survive the journey through Earth's atmosphere.

Halley's Comet

Every 76 years or so, our most celebrated comet returns for another visit. Halley has been documented by humans on each of its last 30 visits, over more than two millennia. Halley's Comet was named for Edmund Halley, the English astronomer who first suggested that comets chronicled in 1531, 1607 and 1682 were actually the same celestial body. The comet was named for him when his prediction that it would return between 1758 and 1759 came true.

After Halley's momentous visit in 1910, when it passed within 0.15 AU of Earth, scientists eagerly prepared for its

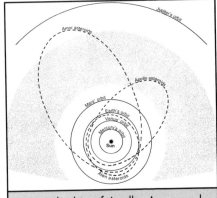

Typical orbits of Apollo, Amor, and Aten asteroids.

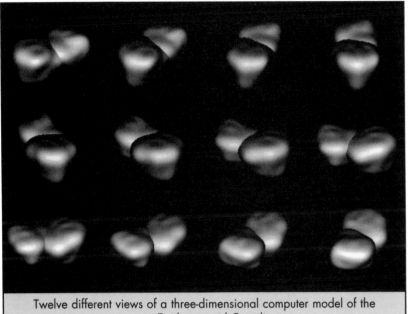

Twelve different views of a three-dimensional computer model of the near-Earth asteroid Castalia.

return in February 1986. Unfortunately for us, naked-eye viewing from Earth was poor in 1986 because the comet passed nearest the Sun while almost exactly on the side of the Sun opposite Earth. Fortunately, Soviet, Japanese and European space agency spacecraft intercepted the comet as it neared Earth. For the first time, scientists were able to take detailed measurements and photographs of a comet's nucleus. Sensors revealed that water ice makes up a much larger proportion of the nucleus than anyone had thought, almost 80 percent.

Most incredible was the photo taken by the European Space Agency's craft *Giotto* (see illustration p.40). Dust pummeled the spacecraft, causing the ship to rotate away from the comet's nucleus, but fortunately not before this phenomenal photo was taken only 1,930 kilometers from the nucleus (*Giotto* approached within 600 kilometers). The nucleus is elongate, approximately 15 kilometers long and 8 kilometers wide, and is much blacker than scientists suspected it would be. Bright jets of light emanate from sublimating gases under the nucleus' surface.

Halley is the only comet that exhibits spectacular cometary phenomena yet occupies a predictable orbit. Fortunately for us, in the long-distant past when Halley became a short-period comet, it got past Saturn and Jupiter, then settled into an orbit that was immune to the heat of the Sun and to the perturbations caused by the large planets. Halley never gets close enough to the Sun to burn up or to be torn apart by the immense gravitational forces. Although Halley ventures from a perihelion (its closest point to the Sun) of 0.59 AU (inside Venus' orbit) to an aphelion (its farthest point from the Sun) of 35 AU (beyond Neptune's), it returns within a single human lifetime. The comet is unusually large, resulting in exceptional displays.

Near-Earth Objects

The greatest threats to our planet come from the short-period comets and near-Earth asteroids, collectively called *near-Earth objects*, or *NEOs*, that share our inner solar system. More than 200 NEOs, ranging in length from 40 kilometers to 10 meters, have been identified. Astronomers estimate that between 5,000 and 10,000 NEOs with diameters of 500 meters or more await discovery and there may be 2,000 NEOs larger than one kilometer. Because NEOs are small and dim, they have only recently been detected.

Near-Earth asteroids are fragments of main-belt asteroids that wind up in the inner solar system. There are three types, defined by their orbits. The Atens orbit very close to the Sun, with a mean distance to the Sun smaller than one astronomical unit and an aphelion distance greater than 0.983 AU. The Apollos orbit just beyond the Earth, with a mean distance to the Sun equal to or larger than 1 AU and a perihelion distance less than or equal to 1.017 AU. The Amors, which travel between Mars and Earth, have a perihelion distance between 1.017 and 1.3 AU. Near-Earth asteroid orbits are highly chaotic and some cross our planet's path.

Most NEOs travel close to Earth but do not ever cross Earth's orbit; of the known Earth-crossing objects, 26 are short-period comets. They are difficult to locate because they reflect little light. The largest Earth-crossing asteroids catalogued are 1627 Ivar and 1580 Betulia, each with a diameter of about eight kilometers. The census is complete for all NEOs eight kilometers or more in diameter; however, scientists estimate that less than 10 percent of the objects whose diameter is one kilometer or less have been discovered. There may also be many with diameters between three and five kilometers in size still to be found. Each could strike a devastating blow to Earth. Three small asteroids, 1981 Midas, 1915 Quetzalcoatl, and 2201 Oljato, will intersect Earth's orbit (but will not strike Earth) before the end of the century.

There is mounting evidence that many near-Earth asteroids are actually binary or multiple bodies. Recent radar studies of 4769 Castalia indicate that it is strongly bifurcated; it is two one-kilometer-sized asteroids that fused in a gentle collision. About 25 percent of radar-detected near-Earth asteroids show some bimodal distribution in their signal but it is not enough to determine whether or not they are truly binary. Multiple asteroids can pull apart in a planet's gravitational field and each piece can eventually strike the planet.

NEOs have short existences compared to main-belt asteroids and long-period comets, which live in their distant habitats for billions of years. NEOs are thought to have a life-span of between 10 million and 100 million years. Where do they come from and where do they go?

Computer models show that about 20 percent of all NEOs fall into the Sun. If an NEO ventures towards an inner planet, the planet's gravity may draw it in or fling it from the solar system. A new supply of near-Earth asteroids comes from main-belt asteroids that are knocked from their orbits by impacts or shoved from their orbits by encounters with Jupiter. Scientists estimate that about 100 fragments larger than one kilometer are added to near-Earth orbits every million years due to disruptive impacts in the main belt. Extinct comets, ones that have depleted their volatile supply, are also candidates for capture into a near-Earth orbit; it is estimated that one-third to one-half of all near-Earth asteroids are extinct comets. Comets can become near-Earth objects when they are perturbed from their locations in the Kuiper belt or the Oort cloud.

Meteorites—
They Strike Planets, Don't They?

O nly recently have people accepted the notion that rocks could plummet through space and strike the Earth. On hearing the news that several stones fell to the ground after a fireball exploded over Weston, Connecticut, on December 14, 1807, President Thomas Jefferson responded, "I would more easily believe that two Yankee professors would lie than that stones would fall from heaven." We now know that meteorite strikes are fairly common.

Meteorites are broken chunks of celestial bodies—comets, asteroids and even planets. For a rock to find its way from the asteroid belt, the Oort cloud, the Kuiper belt or the surfaces of the Moon or Mars to the small portion of space occupied by Earth's orbit, is a difficult undertaking. Surviving the descent through the Earth's atmosphere to its surface is nearly impossible. Yet throughout Earth history meteorites have landed here. They give scientists clues about their origins and their journeys to Earth.

Approach and Landing

A small asteroid or a dead comet that enters interplanetary space is called a *meteoroid*. A meteoroid that penetrates Earth's atmosphere is a *meteor*. A meteor that reaches Earth's surface is a *meteorite*. Each day many small meteorites bombard our planet.

It is highly unlikely that an asteroid or comet will become a meteorite. First, it must enter the inner solar system and be pulled into an Earth-intersecting orbit. If it makes it into the Earth's atmosphere, it must

Meteorite streaking across the night sky.

stand up to the intense frictional heating that melts and vaporizes its surface. Small meteors fizzle out. Larger meteors break into pieces that later disintegrate. Composition also plays a role in the survival of a meteorite; metallic meteorites are far more likely to reach the ground intact than rocky meteorites.

Earth's largest meteorite Hoba in Namibia.

But what a show! Before meteors vanish, they create such brilliant spectacles that they have long been known as "shooting stars" or "fireballs". Scientists refer to exploding meteors as "bolides". Speeding through the atmosphere at more than 40 kilometers per second, a meteor's external temperature can reach more than 1,600° C. Between 100 and 150 kilometers above the Earth, material from its surface turns to liquid, then gas, and becomes incandescent. Although the rock may be only one or two feet in diameter, the bright air mass surrounding it may be hundreds of feet across. A bolide is visible for hundreds of kilometers and is often brightly colored. Normally, the rock disintegrates after only a few seconds, but its dust trail may endure for many minutes. Crackling or hissing sounds, even sonic booms, accompany the light show. Meteorites are often cone-shaped, since material from the projectile's nose vaporizes preferentially. Heating during atmospheric flight also causes the meteorite's surface to melt, forming a black fusion crust.

Moving the 30 metric ton meteorite Ahrighito to its new home in the American Museum of Natural History in New York City.

Meteors fly through the sky individually or in clusters. Individual shooting stars are not as rare as you might think. On a clear Moonless night, three or four may be visible each hour, but their occurrence and their

Meteorite being retrieved from Antarctic ice.

paths are not predictable. More predictable are meteor showers. These fireballs are solid fragments of comet tails that continue to trail behind the comet long after it has gone. Several times a year stargazers gather to witness a well-known event. During a shower between one dozen and several dozen shooting stars can be spotted in a single hour.

Rare Survivors

The largest known meteorite rests where it fell, in a two-meter depression in Namibia, Southwest Africa. Like all great meteorites, Hoba, as it is named, is a chunk of metallic iron and nickel, called an *iron* meteorite. It measures 2.7 by 2.7 by 1 meters and weighs more than 54 metric tons. Hoba was discovered in 1920 but as it is highly weathered by rain and wind it has probably been on Earth for a very long time.

Another enormous meteorite is the 30-metric-ton Anhingito, the largest of the Cape York meteorites of Greenland. Eskimos once made tools from metal they chipped from it. With the help of hydraulic jacks, ice floats and a huge ice-breaking ship, Arctic explorer Robert E. Peary delivered Anhingito to its current resting place in the American Museum of Natural History in New York City in 1897. Perry used the $40,000 he was paid to finance his successful pioneering expedition to the North Pole.

The Willamette meteorite, a 14-metric-ton iron, is the largest meteorite ever located in the United States. It also resides at the American Museum of Natural History, although in 1990 children of the state of Oregon where it was found mounted an unsuccessful campaign to have it returned.

Microtektites.

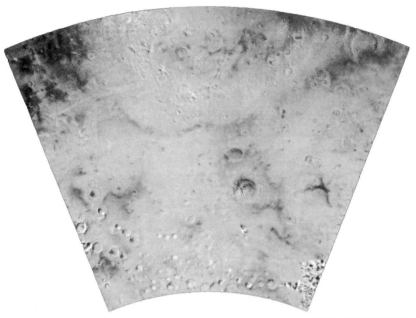

Recognizing an enormous meteorite, especially one made of metal, as being very unusual is not difficult. Almost anyone who stumbled on these rocks would know they were quite extraordinary. But what of the small meteorites or the ones that resemble normal Earth rocks? How are they located?

Photo of Crater Lowell on Mars.

Meteorite collections contain both falls and finds. Falls are actually observed plummeting through the sky and are later located by taking the testimony of the witnesses or by tracking the damage they have caused—hunting methods that are not foolproof. While it is estimated that 560 meteorites weighing 100 grams or more fall within every million square kilometers of Earth's surface each year, only six are found in any given year.

Most meteorites have not been observed falling from the sky so most samples are finds that are collected from the ground. Irons are easily distinguished from Earth rocks but are sometimes too small to be noticed. Stony meteorites, no matter how large, often blend into the background. For this reason, a disproportionate number of irons exist

Chondritic meteorite showing black fusion crust.

in collections. To maximize their chances of finding their bounty, meteorite hunters search in locations with limited vegetation and slow weathering, especially deserts. One or two dozen finds are collected each year.

The most bountiful meteorite repository on Earth is Antarctica. In 1969, a party of Japanese scientists, in Antarctica to study glaciers, found an unprecedented nine meteorites, all of different types. Since then, more than 10,000 individual specimens, many from the same meteorite showers, have been scraped from the Antarctic ice. The moving ice piles up the meteorites at specific junctures and the total lack of vegetation and the acres of surrounding white make the rocks very easy to spot.

Micrometeorites

The solar system is full of microscopic cosmic spherules and interplanetary dust particles. Like their larger siblings, micrometeorites are fragments of asteroids, comets and planets released during cosmic collisions. Micrometeorites that survive a voyage through the Earth's atmosphere are found in ocean sediments and Antarctic ice. On a body with no atmosphere, cosmic dust particles achieve incredible speeds and produce tiny impact craters.

Meteorites from Mars and the Moon

Several known meteorites are too young to have come from the normal sources. Four fell in recent historic times—in France in 1815, in India in 1865, in Egypt in 1911, and in Nigeria in 1962. Six more were found on the ground. These meteorites are 1.3 billion years old, yet asteroids and comets have been geologically dead for more than four billion years. Even the youngest lunar rocks are too ancient to have been the source of these meteorites.

To explain the source of these meteorites, scientists needed to find a body that had enough internal heat to remain active until relatively recently. Only one celestial orb is the right age and is near enough to Earth to be a source of meteorites— Mars. Gas trapped in bubbles in at least two of the meteorites, the fall from Nigeria and one find from Antarctica, match the atmospheric composition of Mars as determined by the *Viking* spacecraft in 1976.

How did the Martian meteorites get to Earth? Researchers point to the extraordinarily large impact craters found on the Red Planet's surface and suggest that at one time an asteroid more than eight kilometers in diameter struck the planet traveling at least 30 kilometers per second. It created a crater more than 170 kilometers across and shot rocks away from the Martian surface, out of the planet's atmosphere, and into an orbit that intersected Earth, about 75 million kilometers away.

Using sophisticated analyses, scientists have also determined that several meteorites found on Earth are in the same chemical and age range as Apollo lunar samples. These specimens are fragments of the Moon that collided with Earth, probably after a large asteroid struck the Moon.

Gifts from the Cosmos

To scientists, meteorites are gifts from the cosmos. How else could they hold fragments of Mars or unaltered pieces of the early solar system in their hands? How else could they study and analyze material that resembles the Earth's core or lower mantle, or age-date rocks to chronicle events that took place billions of years ago? Meteorites are grouped into two main types—chondrites and achondrites. Chondrites are undifferentiated stones; that is, their parent bodies did not separate into the core, mantle and crust structure of most planets and large asteroids. Because they are undifferentiated, scientists suggest that chondrites are

above:
Iron meteorite.
left:
Stony-iron meteorite showing olivine crystals in iron matrix.

composed of primitive planetary material. They are the stuff from which the Earth formed.

Roughly 90 known meteorites are members of the very special subtype, carbonaceous chondrites. These fragile meteorites contain organic compounds of carbon, hydrogen, oxygen and nitrogen—the molecules that make up living things. Some scientists think that carbonaceous chondrites seeded the Earth with the amino acids and other compounds necessary for the origin of life.

Achondrites come from differentiated parent bodies and include irons, stony-irons and some stones. These meteorites represent the core, mantle and/or crust of the parent bodies that spawned them. Iron meteorites are chunks of an asteroid's core. Stony-irons are beautiful but rare. When sliced open, they display a mixture of the nickel-iron metal found in iron meteorites and the mineral olivine, common in stony meteorites. Scientists suspect that these meteorites come from near the core-mantle boundary of the parent body, a place where liquid metal from the core mixes with mantle crystals.

Clocks in Rocks

How do scientists determine the timing of events that took place in the distant past? How do they calculate the age of a meteorite or rock? What evidence do geologists need to determine that a meteorite must have come from Mars?

Fortunately, all rocks include internal clocks—radioactive isotopes. Scientists read these clocks with radioactive-age-dating techniques. The nucleus of an atom contains positively charged protons and uncharged neutrons. Each element has a fixed number of protons but can have different numbers of neutrons. For example, a carbon atom's nucleus must have six protons, but may have anywhere from three to ten neutrons. Carbon's isotopes are named for the total number of protons plus neutrons contained in its nucleus, for example, carbon-9, carbon-10, etc.

The nuclei of many isotopes are unstable. To achieve stability, a nucleus ejects a particle; this process is known as radioactive decay. In a sort of cosmic alchemy, an isotope of one element becomes an isotope of another. For example, carbon-14 changes to nitrogen-14 when one of its neutrons emits a tiny negative particle (a negatively-charged electron or beta particle) and an antineutrino (an uncharged particle containing a small amount of kinetic energy) and becomes a proton. The element nitrogen always has seven protons.

Although it is impossible to determine when a particular isotope will undergo radioactive decay, statistically the rate at which this conversion takes place is constant and well known. The amount of time necessary for half of a given group of isotopes to convert to another kind of isotope is referred to as its half-life. The half-life for the transformation of carbon-14 to nitrogen-14 is 5,730 years.

What if you had a pile of 100 carbon-14 atoms for 5,730 years? At that point, only 50 would remain, but you would now also have 50 nitrogen-14 atoms. In another 5,730 years, there would be 25 carbon-14 and 75 nitrogen-14 atoms. Finally, in 573,000 years, or 100 half-lives, your pile would contain 100 nitrogen-14 atoms and no carbon-14 atoms. Suppose there was no nitrogen-14 in your initial pile. At any time during those 573,000 years, you could count the number of carbon-14 and nitrogen-14 atoms and know, within a reasonable margin of error, how many years had elapsed since the pile contained only carbon-14 atoms. This example illustrates how radioactive age-dating, specifically carbon-14 dating, works.

As the years (and half-lives) pass, the number of atoms of carbon-14 decreases and becomes difficult to measure, so this method of dating is useful only for young rocks. If you wanted to determine the age of the Earth, known to be around 4.5 billion years, your carbon-14 clock would be useless. Fortunately, meteorites and rocks contain many radioactive isotopes with half-lives ranging from a fraction of a second to several billion years. A few important isotopes common in rocks and minerals and their half-lives: rubidium-87 decays to strontium-87 with a half-life of 5.0 billion years; potassium-40 decays to argon-40 with a half-life of 130 million years; and uranium-238 decays to lead-206 with a half-life of 450 million years.

Radioactive clocks can be used only to date closure episodes, times when the substance became impermeable to the isotopes. Since living organisms do not collect nitrogen-14, the radioactive clock is set at the moment of death. Scientists can easily date events, such as the cooling of a volcanic rock. Such ages are used to determine the timing of major events on the Earth or in the solar system.

Earth's
Ancient Wounds

he Moon's impact history is written all over its face. With no atmosphere to protect its surface, meteorites of all sizes leave their mark. With no geologic processes to destroy them, meteorite craters, sometimes called *astroblemes* (star wounds), are virtually frozen for all time.

Although Earth is protected from some impacts by its gaseous atmosphere, meteorites have collided with the planet throughout its history. Since about 70 percent of our planet is covered by oceans, it is safe to assume that many craters are buried beneath water and sediment. Plate tectonic processes, including the generation of new and the destruction of old ocean crust and the collisions of continents, have erased most of the remaining wounds from our planet's surface.

In spite of all this, Earth has craters. Some are easily observed on the ground and others are seen only in the distant views afforded by satellites. Sometimes no crater remains after an impact and the prehistoric damage is identifiable only by the characteristic features of impact shock.

Our planet has a long history of bombardment. Earth's scattered craters and other impact scars are the physical memories that remain to tell the tale.

Impact Craters

Craters are formed by two methods—impacts and volcanic eruptions. It was not until the 1950s that scientists accepted that impact craters could be found on Earth. Because the only meteorites ever seen falling to the ground were small, geologists attributed all craters to a volcanic origin. In fact, it took a long time for scientists to embrace the idea that the craters of the Moon were caused by meteorite impacts. Until the *Apollo* missions settled the question, some geologists thought all lunar craters were volcanic.

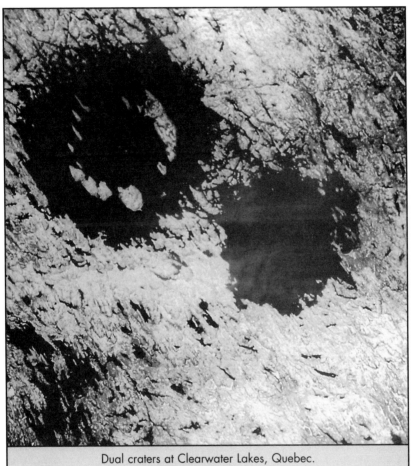

Dual craters at Clearwater Lakes, Quebec.

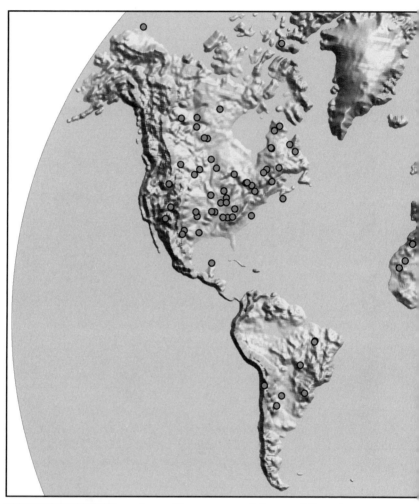

Map showing the locations of Earth's major craters.

According to one of the world's primary crater hunters, Richard A. F. Grieve of the Canadian Geological Survey, 147 impact craters have now been identified on Earth; three to five more are added to the list each year. All of the astroblemes are associated with meteoritic material or impact features from the shock wave that accom-

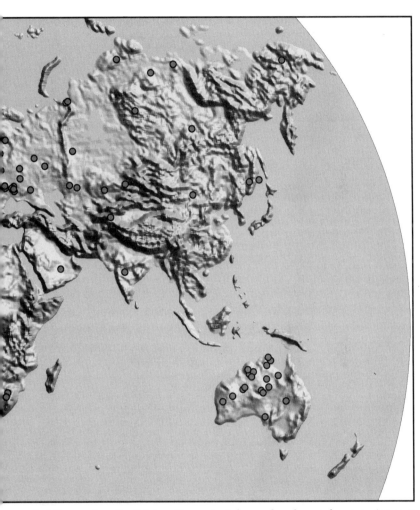

panied the meteorite. Meteorite material may be absent from an impact site either because the meteorite vaporized before it hit the ground or because it weathered away on the Earth's surface. Small craters are more likely to be associated with meteorite material than large ones. Because the velocities of the larger bodies are unimpeded by the atmosphere, large bodies cease to exist as physical entities on impact. Their shock waves are so powerful that in many instances the meteor disintegrates in the atmosphere and the crater and other damage is produced by the shock wave alone.

About two-thirds of the known craters are located on old, interior parts of the continents, far from the boundaries of the crustal plates where plate tectonic activity is focused. These ancient regions, called cratons, are ordinarily hundreds of millions or billions of years old. Cratons are generally flat and geologically stable places where craters and other features are not easily destroyed by geological processes. An astrobleme will be better preserved on a craton than anyplace else on Earth.

Even with the craton's geological stability for protection, the average lunar crater is much older than the oldest Earth crater. Our two oldest astroblemes are the Sudbury structure in Ontario, Canada, and the Vredefort Ring in South Africa. Both are just under two billion years old.

Not coincidentally, Sudbury and Vredefort are also two of Earth's largest craters, 200 kilometers and 140 kilometers across, respectively. Chicxulub Crater, Mexico, measuring 180 kilometers, has been implicated in the dinosaur extinctions 65 million years ago, and the Popigai, Russia and Manicouagan, Canada craters, both 100 kilometers across, may be linked to other extinction events.

Aerial photograph of 3.8 kilometer diameter Brent Crater, Ontario, Canada.

Several of Earth's craters are just a few thousand years old. All of these are small, less than 0.5 kilometers in diameter, but they can be found around the globe, in Saudi Arabia, Russia, Australia, Estonia, Argentina, and even in Haviland, Kansas. The youngest crater is the tiny Sikhote-Alin in Siberia (27 meters), formed in 1947.

The Canadian craton is home to many interesting and ancient craters. Most have been identified using satellite photos, since they are so old that weathering has all but wiped out their surface features. The craters are always surrounded by evidence of shock but never by meteorites since both stones and irons weather rapidly in the harsh Canadian climate. On the continental shelf of Nova Scotia, Canada, lies the only underwater impact crater ever identified (although many more surely exist in the deep ocean). It is the Montagnis crater, 45-kilometers-diameter and 50 million years old.

Although we see few astroblemes on Earth, the planet's cratering rate should be roughly equivalent to the rate at which meteorites strike. In fact, cratering rate is one of the many bits of evidence that scientists use to calculate the Earth's impact rate. Craters smaller than 20 kilometers in diameter are generated less often than might be expected because small meteors break up in the Earth's atmosphere. It is estimated that eleven craters 10 kilometers across or larger are created every one million years, but on average only three of them are on land. Between one and three of these craters are greater than 20 kilometers in diameter. Huge craters, greater than 100 kilometers in diameter, form roughly twice every 100 million years.

Impact Cratering

The abrupt termination of a meteorite's flight by collision with Earth releases vast quantities of kinetic energy, energy that is transformed into pressure and heat. The amount of energy released by the collision depends on the size and speed of the flying object.

The atmosphere plays a large role in how much bang a meteorite has—it can cause the meteorite to fragment, reducing each piece's mass, or it can slow it down through friction. On Earth, enormous projectiles, weighing more than 1,000 metric tons, soar unimpeded through Earth's atmosphere and strike the planet with vicious force. Smaller objects, less than 200 kilograms, are slowed by the atmosphere and pack a far

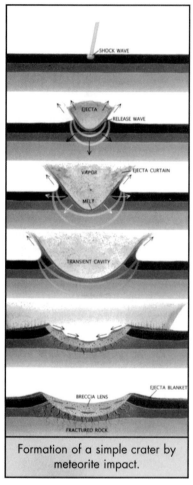

Formation of a simple crater by meteorite impact.

less violent punch. Still smaller meteors break up in the Earth's atmosphere and are unlikely to produce damage. It has been estimated that if the Earth had no atmosphere it would have accumulated about 200,000 impact craters greater than one kilometer in diameter.

The lightning-fast transfer of kinetic energy from the moving meteorite to the underlying rock produces a crater. In the formation of a simple crater, the shock wave compresses the rock, then a release wave springs the earth back in the opposite direction, propelling some of the impacted rock upward and outward. Both the compression and ejection of huge amounts of material create a cavity. Cracks are also induced in the host rock, causing the crater walls to collapse and partially fill in the hole. Simple craters are bowl-shaped and have upraised rims. The impact of a larger meteorite causes the formation of a complex crater. Complex craters are larger than about four kilometers in diameter and are distinguished by a small hill rising in the center of the basin.

Blast Magnitude

The most significant property of a meteorite strike is not its diameter or mass but the force with which it hits. Force is not proportional to size because the meteorite's density, related to its composition, plays an important role in its punch. Force is proportional to kinetic energy, which is typically one hundred times greater than an equivalent mass of TNT. So kinetic energy is expressed in megatons (MT) of TNT.

The nuclear bomb that hit Hiroshima in 1945 was only one one-hundredth of a megaton. Keep in mind, though, that much of the Hiroshima bomb damage was done by radiation and radioactive fallout; a meteorite strike is only equivalent to a nuclear bomb in the amount of energy released.

Photomicrograph of shocked quartz. Field of view is 1.5 mm.

Shock

The shock effects of impacts are caused by the immense temperatures and pressures generated by the meteorite and its shock wave at the moment of impact. Most shock effects are microscopic, recognized in individual minerals. One important shock feature, shocked quartz, forms when quartz, one of the Earth's most abundant minerals, is subjected to high pressure. Shocked quartz has much higher density than normal quartz due to the effect of pressure on its structure. Another shock feature, tektites, are formed when the vaporized meteorite and host rock material recondenses into glassy spheres and falls back to Earth.

Shatter cone.

Shatter cones are the only macroscopic structures associated with shock. Shatter cones rise between a centimeter and two meters above the ground and have radial fractures extending outward from their apex. Common where meteorites strike limestone or other fine-grained rock, they form when a shock wave fractures nearby rocks into a cone shape. Interestingly, their noses always point toward the impact site.

Meteor Crater, Arizona

Meteor Crater, Arizona, was the first impact crater to be recognized as such on Earth. The 1,200-meter-wide astrobleme lies in the high desert 80 kilometers east of Flagstaff. The surrounding region is flat, held up by the limestone that forms the top of the Grand Canyon. Although the nearest volcanoes are 30 kilometers away, early visitors attributed the 170-meter-deep crater to volcanic activity. It seems impossible that naturalists could have made the mistake because the crater is surrounded by jagged-edged iron fragments whose only known source is iron meteorites. Yet they disregarded the evidence, suggesting that it was a coincidence.

In 1903, Daniel M. Barringer defied the common assumption that the crater was volcanic and decided that the iron fragments and the crater were related. He calculated that the enormous chasm was created by a gigantic meteorite, containing millions of tons of iron and hundreds of thousands of tons of nickel. Although he had never seen Meteor Crater, he staked a mining claim immediately.

Barringer drilled holes into the crater for years, but found no evidence of volcanism. But he was perplexed because, beyond a layer of broken rock and some oxidized iron, he detected no indications of a large meteorite either. Loss of his investors' confidence and the onset of the Great Depression shut down Barringer's mining operation. Later calculations showed that an iron meteorite of the size necessary to create a 1.2-kilometer-wide crater would travel through Earth's atmosphere unaffected and would vaporize on impact. The crater, also known as Barringer's Crater, is still owned by his family and is a major tourist attraction.

The lack of a meteorite fueled the argument of those who believed in a volcanic origin for the crater. It was not until 1957 that a young geology graduate student named Eugene Shoemaker convinced scientists that the iron fragments, the layer of broken rock, and the shock-melted glass tektites could only have formed as a result of a meteorite impact. The Earth had its first official meteorite crater. Because Meteor Crater is a young and well-preserved impact crater, it is still used by geologists wishing to study bombardment features. In the 1960s, Apollo astronauts took field trips to the Crater to prepare them for examining the geology of the Moon.

Over the years the creation story of Meteor Crater has unfolded. Between 20,000 and 50,000 years ago, an iron meteorite, 60 meters in diameter, entered the Earth's atmosphere. Friction did not retard its 25-kilometers-per-second speed, but it broke it into thousands of small pieces, which scattered across the land at the time of impact. The few large chunks became superheated and vaporized. The attendant shock wave so compressed the host rock that it became hot and flowed under the extreme pressure. The vaporized meteorite and much of the host rock blasted from the crater, causing the famous northern Arizona rock layers to tilt backwards and forming a 330-meter-deep simple crater. Vaporized meteorite and rock debris rained down and partly filled the hole. The total time from impact through the development of the crater was less than one second. The explosion was like a 1.7 MT bomb, equivalent to the most powerful nuclear devices.

Artist's conception of the birth of Meteor Crater, Arizona.

CHAPTER SEVEN

Demise
of the Dinosaurs

inosaurs ruled the world for 160 million years, a long time compared with the 400,000 years our own species has been on the planet. Dinosaurs were agile, cunning and, many paleontologists believe, warm-blooded. They were well-adapted to a variety of environments and were the reigning members of each community of animals and plants, or ecosystem, they inhabited. As a group, they were quite diverse. Most were docile plant eaters but some, most notably the villain Tyrannosaurus rex, were the most vicious flesh-eaters the world has ever seen.

How could creatures of such diversity and longevity have been so thoroughly erased from the planet? That question has been the source of controversy for decades. The most popular, and the most likely, theory involves a cosmic collision. One day, 65 million years ago a giant meteorite struck Earth, beginning a chain of events that extinguished not only the dinosaurs but two-thirds of the planet's other species.

Geologic Time

If we compressed the 4.5 billion years of Earth history into a 24-hour day, the interval since the evolution of modern humans would be represented by only 7.7 seconds. By contrast, the 160-million-year reign of the dinosaurs would be equal to 51 minutes.

To handle the enormousness of Earth's time frame, nineteenth-century naturalists developed the geologic time scale. Its divisions were based on major events in Earth history, most of them periods when many species of animals and plants disappeared, called mass extinctions. For example, the last period of dinosaur dominance is called the Cretaceous (identified by the symbol "K"); fossils in Cretaceous rock strata are typified by dinosaurs, conifer tree pollen, and small simple mammals. Rocks

> Geologic time scale.

TIME SCALE	ERAS	DURATION OF PERIODS	PERIODS			DOMINANT ANIMAL LIFE
	CENOZOIC 70 MILLION YEARS DURATION		Quaternary		Recent Pleistocene	Man
10 20 30 40 50 60 70		70	Tertiary	EPOCHS	Pliocene Miocene Oligocene Eocene Paleocene	Mammals
80 90 100	**MESOZOIC** 120 MILLION YEARS DURATION	50	Cretaceous			
150		35	Jurassic			
		35	Triassic			Dinosaurs
200		25	Permian			
		20	Pennsylvanian			
250		30	Mississippian			Primitive Reptiles
300		65	Devonian			Amphibians
350	**PALEOZOIC** 350 MILLION YEARS DURATION	35	Silurian			Fishes
400		75	Ordovician			
450 500		90	Cambrian			Invertebrates
Figures in millions of years	PROTEROZOIC ARCHAEOZOIC	Figures in millions of years	1500 MILLION YEARS DURATION			BEGINNINGS OF LIFE

of the following period, the Tertiary (symbolized by a "T"), contain no dinosaur fossils, but include a profusion of mammal and flowering plant species. The transition between the two is the Cretaceous-Tertiary (K-T) boundary. The K-T boundary also separates two broader categories of time—the Mesozoic Era (meaning middle life), when reptiles dominated the planet, and the current Cenozoic Era, (meaning new life), the age of mammals. We live in the Holocene Epoch, which is part of the Quaternary Period of that Cenozoic Era.

Mass Extinctions

Species extinctions are not uncommon over geologic time; several species vanish every few million years. During these relatively stable periods, evolution mostly takes place by the gradual processes of natural selection; that is, by competition and survival of the fittest. In natural selection, the organism that competes more successfully for resources and mates leaves more offspring, guaranteeing its evolutionary success.

Times when many more species vanish than normal are called mass extinctions. Scientific debate rages over whether mass extinctions are caused by gradual environmental changes or more drastic occurrences. In either case, after a mass extinction, it is common for the surviving organisms to diversify rapidly and take over all the vacant ecological roles. Therefore, animal and plant communities before and after the extinction event are very different populations.

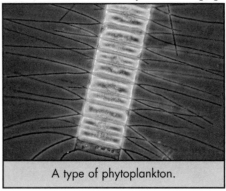

A type of phytoplankton.

Paleontologists, the scientists who study fossils, recognize 24 noteworthy extinction events, 5 major and 19 minor, over the last 540 million years. In each of these periods, between 25 and 95 percent of species perished. The most recent major extinction event was the one that took place 65 million years ago, at the K-T boundary.

Fossil ammonites.

Dinosaurs Didn't Go Alone

While dinosaurs were the most famous organisms to become extinct at the end of the Cretaceous, the tragedy was far more widespread—between 65 and 75 percent of all of Earth's organisms vanished. Hardest hit were the land animals. Scientists estimate that no species of land animal weighing more than 25 kilograms (55 pounds) survived into the Tertiary. The small reptiles and amphibians who dwelled near lakes and streams endured, as did the insects. Only a few species of our rodent mammalian ancestors survived the devastation. In all, although 88 percent of land-dwelling species vanished, as many as 90 percent of those inhabiting fresh water survived.

Land plants suffered greatly, although the species whose seeds can linger in the soil before germinating recovered rapidly. As a result, there was an explosion in the abundance and diversity of flowering plants in the Tertiary.

Marine organisms were not spared—almost 50 percent of marine species died off. As many as 95 percent of the phytoplankton species, the tiny photosynthesizing plants that form the basis of the marine food chain, vanished. Phytoplankton are eaten by small animals which in turn are eaten by larger ones, so when the phytoplankton perished, the food distribution structure of the ocean collapsed. Larger marine animals, such as the beautiful spiral-shelled ammonites, died abruptly. Only creatures of the deep sea entered the Tertiary in relatively good shape. Many scientists believe that the numbers of some species may have been reduced before the end of the Cretaceous, however.

Without question, the K-T extinctions were a global disaster of unimaginable proportions. Any theory that attempts to explain them must account for the demise of the tiny marine plankton as well as the giant land-dwelling dinosaurs. It must also explain why other organisms, most notably the freshwater animals, survived.

Cretaceous-Tertiary boundary clay layer near Gubbio, Italy.

Agents of Extinction

For decades, theories about dinosaur extinction invoked gradual environmental changes—the oversized reptiles could not evolve fast enough to keep up with a cooling climate, a lowering sea level, or a change in vegetation. One popular notion was that the lumbering hulks were outwitted by their smaller and smarter mammalian competitors—possibly the mammals ate all the reptiles' eggs. Many of these theories fall short of explaining K-T extinctions because they disregard the disappearance of so many other species. Moreover, the notion of the cold-blooded, dim-witted dinosaur is a thing of the past—scientists now think dinosaurs were warm-blooded, smart and quick.

Scientists also advanced the idea that large volcanic eruptions were common in the late Cretaceous. Organisms were extinguished at the K-T boundary by environmental changes due to the poisonous dust and gases released into the atmosphere by volcanoes. Some scientists even suggested that the volcanic dust carried toxic compounds that reduced dinosaur fertility.

In 1980 a completely different type of theory was advanced—one that involved an extraterrestrial agent. Although the idea had been suggested and discarded previously for lack of evidence, compelling new data was discovered to support the concept that the dinosaurs and their contemporaries were extinguished as a result of a cosmic collision. Since the theory was advanced, scientists have found much corroborative evidence and have calculated many insidious ways in which life could have been destroyed in the aftermath of the collision.

Birth of the Impact Theory

When geologist Walter Alvarez travelled to Italy to study the sedimentary

Diagram of iridium concentration in Gubbio, Italy, sediments showing spike at Cretaceous-Tertiary boundary.

rocks of Gubbio he was intrigued by a one-centimeter-thick clay layer that fell at the K-T boundary. He hypothesized that the clay represented a short time span, geologically speaking, so he went to his father, Nobel prize-winning physicist Luis Alvarez, with a question. Is there any way to calculate the length of time necessary for such a clay layer to be deposited?

The senior Alvarez had no experience with the earth sciences, but he quickly devised a plan. He knew that the abundance of platinum group elements in surface rocks was minuscule. For example, iridium concentration is only about 0.03 parts per billion at the Earth's surface. Iridium that was present at the time the planets accreted followed iron into the core during planetary differentiation. The iridium in sediments originates in the constant rain of cosmic dust particles. The rate of the dust-particle rain should not have changed in at least a few billion years. Therefore, Alvarez reasoned, the quantity of the element in the soil layer might help one calculate the time required for deposition of that layer.

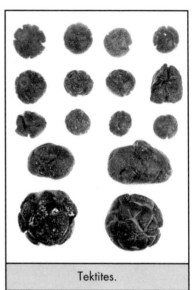

Tektites.

The Alvarez team was joined by Frank Asaro and Helen V. Michel who analyzed iridium in the Gubbio clay. The results were astonishing. While the levels of iridium in the limestone beds that surrounded the clay layer were normal, the clay itself had a concentration of about 3 parts per billion, nearly 100 times the expected amount! After repeating the analyses, the scientists determined that the source of the excess iridium could only be extraterrestrial. They concluded that the element had come from a mete-

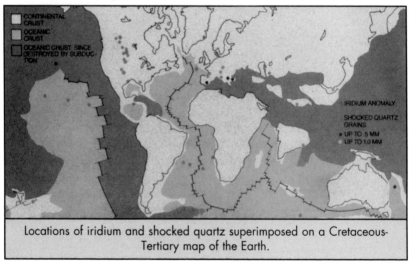

Locations of iridium and shocked quartz superimposed on a Cretaceous-Tertiary map of the Earth.

Boulder in sedimentary deposits created by giant waves caused by Chicxulub impact across Gulf of Mexico.

orite impact, because chondrites, the most common type of meteorites, have iridium concentrations of about 500 parts per billion.

Before going public with their findings, the team measured iridium in other K-T boundary rocks and found even higher values. They calculated that the total world-wide iridium anomaly was about five million kilograms (0.5 million tons). To introduce that much iridium into the Earth's environment would require a meteorite 10 kilometers in diameter, slightly larger than the nucleus of Halley's Comet. Although more recent calculations of the size of the impactor have been carried out, that estimate remains the same.

In their paper, published in *Science* magazine the next year, Alvarez father and son and their colleagues proposed that a giant asteroid or comet had struck the Earth. Plants and animals not killed outright by the impact starved to death because the enormous quantities of dust and gas kicked up by the impact blocked the sunlight for several months, making photosynthesis impossible. As a result, entire food chains collapsed.

A Flurry of Activity

News of the fantastic new theory spread rapidly. Scientists in many fields began to search for information to corroborate or disprove the idea. Researchers have split into three main camps—the impactors, the volcanists and the gradualists. Intense research continues today with scholarly arguments taking place in journals and at scientific meetings, but overwhelming evidence for the impact theory has brought many

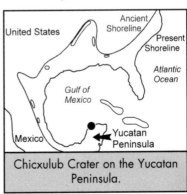

Chicxulub Crater on the Yucatan Peninsula.

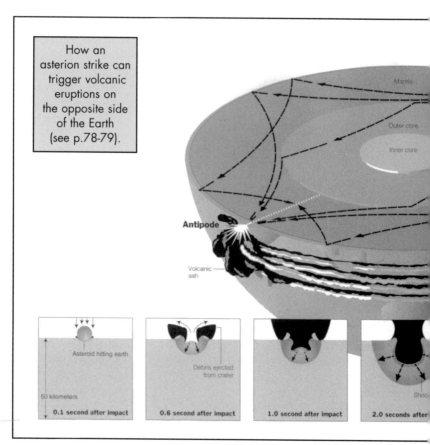

How an asterion strike can trigger volcanic eruptions on the opposite side of the Earth (see p.78-79).

Mantle

Outer core

Inner core

Antipode

Volcanic ash

Asteroid hitting earth

50 kilometers

0.1 second after impact

Debris ejected from crater

0.6 second after impact

1.0 second after impact

Shoc

2.0 seconds after

skeptics around.

After publication of the Alvarez paper, the iridium anomaly was identified in K-T boundary layers around the globe. To date, more than 120 anomalies in marine and terrestrial sediments have been located. Other platinum group elements show the anomalous values expected from meteorite contribution, as well.

Researchers uncovered other signs that support the meteorite impact theory. Shocked quartz, thus far only identified at meteorite craters and nuclear explosion sites, was detected in the K-T boundary clays. Most scientists believe that this form of quartz is formed only at

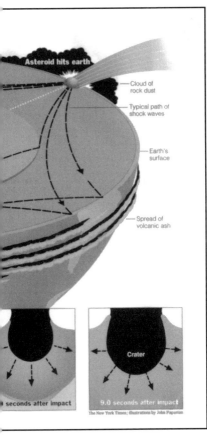

the high pressures associated with impacts. Tektites, the droplets of glass that recondense from vaporized meteorite and host rock in the atmosphere, were also discovered in the 65-million-year-old clay layers. In one stunning find, a half-meter thick layer of tektites was located on the Caribbean island of Haiti.

Scientists also noticed an abundance of soot in K-T sediments from around the world and inferred that it was the by-product of global wildfires. A 350-meter thick K-T-age rock layer containing boulders up to 12 meters across was found on the south coast of Cuba. Geologists concluded that the deposit was the outcome of tsunamis, or massive sea waves, generated by an impact on ocean water.

Because of the thick tektites on Haiti and the tsunami deposits on Cuba, the search for the most important evidence, the impact crater, narrowed in on the Caribbean.

The Smoking Gun

For a decade, members of the impact camp could not point to the site of the impact crater, the evidence they most needed to convince the skeptics. Finally, an enormous crater, called Chicxulub, was identified buried beneath almost two kilometers of sediment on the northern coast of the Yucatan Peninsula of Mexico. The circular hole is at least 180 kilometers wide, the largest collision event we have any record for since life first flourished 570 million years ago. It was identified by petroleum geologists using geophysical techniques to search for oil off of Mexico.

Chicxulub's age is identical to that of the impact debris scattered over the Earth. The debris, including iridium and tektites, can be traced

from the crater around the globe and is used to link Chicxulub to the K-T boundary extinctions. Members of the impact camp believe that this is irrefutable evidence that the meteorite was the sole extinguisher of life.

Although many skeptics were converted by the discovery of the crater, there are still scientists who suggest that the end of Cretaceous extinctions were already underway by environmental or volcanic mechanisms before the meteorite hit, and that the collision played a small role at most.

An Alternative—Killer Volcanoes

At about the time that the impact scenario was gaining favor, a small but vocal group of scientists began working on an old idea, but with a different slant. In this theory the agent of destruction is also a catastrophe, but the origin of the disaster is right here on Earth—volcanoes. Volcanoes are common on the Earth's surface and have been for virtually all of geologic time. But once in a while, the theory goes, volcanic eruptions become so intense that the gases and debris shot high into the atmosphere bring on global environmental changes and mass extinctions.

Earth is home to a unique type of volcanic rock formation, called flood basalts. Flood basalts can cover more than two million cubic kilometers. Geologists

Plume with large head and small tail.

once assumed that to accumulate that much material, the basalts erupted slowly over very long periods of time. But improvements in radioactive age dating techniques revealed that the total duration of most flood basalt eruptions is short, less than one million years.

A new theory for the formation of these volcanic provinces had to be developed. The idea is that excess heat from Earth's core causes material deep in the mantle to rise to the surface. Tank experiments with fluids of different densities show that the warmer and therefore less dense material rises as a plume, with a distinct head and tail structure, through the more dense fluid. Flood basalts are the product of volcanoes that erupt when the plume head of hot material reaches the surface.

Fortunately, there are no plume heads erupting today. To deposit millions of cubic kilometers of material in less than one million years would require an eruption rate more than 100 times greater than the

Deccan Traps, India.

largest eruptions witnessed in the last century. Plume tails do erupt today; they are known as hotspots and are the roots of many of Earth's volcanoes, including those on the Hawaiian Islands.

Of course, for volcanists to successfully blame the K-T extinctions on a flood basalt, there must be one somewhere that is of the right age. And indeed, there is. The Deccan Traps of India are 65 million years old and erupted over a period of less than one million years. Such intense volcanism would introduce millions of tons of sulfur dioxide and other gases into the atmosphere, blocking the sun and causing acid rain. Carbon dioxide from volcanoes would warm the planet through the greenhouse effect.

With the identification of Chicxulub crater, the volcano camp no longer denied that a meteorite impact occurred at the K-T boundary. Their doubts now focus on the impact's importance to the mass extinctions. Some volcanists suggest that the extinction was well under way due to Deccan volcanism and that the subsequent collision did little harm.

You Can Have it Both Ways

The closeness in time between the creation of the second-largest impact crater on Earth and a major extinction event must be more than coincidental. Impact debris is found in the same layer of rock that holds evidence of the extinction of many species; this phenomenon centers on Chicxulub and spreads around the globe. However, the existence of an enormous pile of volcanic rocks also of the right age is evidence for intense volcanism as the agent of the K-T extinctions. How can these conflicting notions be resolved?

An elegant but very controversial new theory combines the impact and volcano scenarios: a cosmic collision may have triggered the flood basalt eruptions. The idea is that when the meteorite struck, seismic waves traveled outward through the Earth from ground zero. The waves sent intense energy to the side of the planet opposite the impact, called the antipode, and generated enormous earthquakes. Some scientists suggest that the impact triggered mantle plumes, which they believe are found in antipodal pairs at opposite sides of the globe. In this scenario, each half of the Earth would be treated to its own disaster—a meteorite impact on one side, volcanic eruptions on the other.

Support for this new theory comes from the planet Mars. Mars' largest impact basin, Hellas Plenitia, is antipodal to the solar system's largest volcano, Alba Patera. Scientists think that fractures at the impact's antipode penetrated more than 15 kilometers deep and helped trigger the lava flows. The antipodal relationship between impact craters and large volcanic eruptions has been identified at other locations in the inner solar system, as well.

Of course, the new theory has problems. Most important is that the Deccan Traps are not antipodal to Chicxulub crater. The Deccan's antipode is in the eastern Pacific Ocean where impact evidence is difficult to uncover. Therefore, this theory requires that a second, probably even larger, meteorite struck the Earth at roughly the same time, a meteorite impact for which we currently have no evidence.

Whether it triggered flood basalt eruptions or not, the evidence is overwhelming that a meteorite was primarily responsible for the K-T extinctions. The collision theory is preferred by most of the scientists involved in K-T research. The next question is, what was it like at that moment between the end of the Cretaceous and the beginning of the Tertiary?

Out with the
Old World, In with
the New

It seems irrefutable that a cosmic collision with a massive meteorite caused widespread environmental destruction and was primarily (although perhaps not solely) responsible for the K-T extinctions. The numbers of some Cretaceous species may have been reduced before Chicxulub hit but, in general, life was flourishing at the moment of impact.

With each new study of K-T boundary rocks, our understanding of the last days of the dinosaurs becomes more terrifying. What began in the Alvarez paper as simple global cooling due to a cloud of impact debris has grown into a horde of potential killers—freezing cold, complete darkness, enormous tidal waves, greenhouse heating, intense acid rain and global wildfires.

Shortly after the beginning of the Tertiary, the environment gradually became less severe. With the dinosaurs gone, survivors multiplied, evolved and took over the vacant habitats. Chief among the new regime were the furry little mammals who were destined to rule the next geologic age, our current time, the Cenozoic.

What about the other mass extinctions that have taken place throughout Earth history? Have they too been caused by extraterrestrial impacts? Might such a tragedy occur again?

The End of the World

To the dinosaur who looked up that ruinous day 65 million years ago, the meteor, 10 kilometers in diameter, would have become visible about three hours before impact. If it was an asteroid, it would have been traveling at about 15 kilometers

> The star of doom.

per second; if it was a comet its speed would have been up to 60 kilometers per second. At those velocities it would have taken only one to 10 seconds for the meteor to travel through the atmosphere, but it would have produced a brilliant fireball. The object may have fragmented, but 10 kilometers is a large fraction of the thickness of the atmosphere, and the pieces would not have spread far.

The explosion hit with 10,000 times the force of today's total world arsenal of nuclear weapons, greater than 100 million MT of energy. Heat energy and the passing shock wave annihilated everything in sight. The meteorite or its shock wave excavated a crater more than 25 kilometers deep. On one side it hit water, initiating tsunamis more than five kilometers high. Intense seismic waves resulted in world-wide earthquakes.

On impact, the meteorite chunks and underlying host rock vaporized. A plume of dust and gas exploded outward, taking some of the atmosphere with it. Earth was stripped of part of its protective gaseous blanket.

For many of Earth's inhabitants, loss of the atmosphere was the least of their problems. Dust and gas coalesced to form fireballs that followed ballistic paths around the planet (which is why the impact debris is globally distributed). Heat energy, released as the material re-entered the atmosphere, was equivalent to the energy put out by a kitchen oven —a world-wide kitchen oven— and animals living above ground may have been broiled. Forests burst into flames; in all, plant matter equal to half the world's current forests was incinerated. The impact and its initial consequences would have caused a mass mortality, but the horrors were mostly local and most species had some survivors.

It was the aftermath of the collision that caused the mass extinction. Impact dust and smoke from global fires blocked sunlight entirely for two to three months and hindered photosynthesis for up to one year. An estimated one kilogram of dust, the first sediments of the Tertiary, fell out over every square meter of Earth's surface within a few months of the impact. Photosynthesizing plankton died and most marine food chains were destroyed. In some land areas, surface temperatures were reduced by 20° C for more than six months. Plant seeds and spores might have survived the unbearable cold in the blankets of fallout dust.

The location of the meteorite impact site assured that the extinctions were as severe as possible. The Yucatan Peninsula is covered by a deep limestone layer, a rare occurrence on Earth's surface. Limestone is made of the shells of small animals and is rich in carbon, sulfur and water vapor. After the impact, oxygen and nitrogen in the atmosphere combined with water vapor from the limestone to produce nitric acid. Also, more than 100 billion tons of sulfur dioxide gas were released by the limestone, forming sulfuric acid. Initially the acidic gases assisted the impact debris in blocking out sunlight but eventually they condensed to fall as a lethal acid rain, killing ocean-dwelling organisms by dissolving their shells. Researchers have proposed that acid rain dissolved some animal bones on land, leading to the barren layer that in some places separates the youngest dinosaur fossils from the K-T boundary.

Carbon released from the limestone combined with atmospheric oxygen to form carbon dioxide (CO_2), the greenhouse gas that is responsible for the global warming we see today. After the impact, atmospheric CO_2 increased two to five times, trapping ultraviolet radiation and resulting in devastating heat. Fluctuations between the extremes of freezing climate caused by debris and global warming from atmospheric CO_2 may have lasted for several years.

The scenario presented above is the result of computer modeling of the impact event and scientists' interpretation of the fossil record. Although it is difficult to know which environmental changes posed the biggest hazard, the end result was devastation to most life. It is likely that an identical impact today would wipe out civilization.

Who Survived?

The key to surviving the destruction was diet. Animals who were linked by their food chain to a diet of living plant tissue were demolished—that's probably what killed off the last of the dinosaurs. Marine animals dependent on phytoplankton were devastated.

It was the food chains rooted in organic debris, known as detritus, that came through the extinctions best. In the aftermath of a global dying, detritus would have been plentiful. The animals that fed on organic matter left behind in lakes, streams, soil and rotting logs, and the animals that in turn fed on them, survived. Even the remaining plankton may have been detritus-feeders.

And what of our ancestors, the mammals? The surviving mammals were small, and not very specialized. They were either insect-eaters or omnivores, able to subsist on whatever they found, such as the insects, spiders, worms and larvae that live on rotting logs and other detritus.

Thank Your Lucky Stars

During the Cretaceous, reptiles dominated every ecosystem while mammals scurried around in the crevices. The dinosaurs, as successful as they were, were not evolving larger brains (which we equate with increased intelligence), but then neither were the mammals. It's been said that the smartest mammal was no smarter than the most intelligent dinosaur. The dominance of reptiles over mammals was stable and probably could have continued for hundreds of millions of years.

It was a random event, and an extraterrestrial one, that wiped the slate clean. Dinosaurs, who were large and well-adapted to their environments, were done in by their size and perhaps by their love of fresh vegetables or the animals that eat them. Mammals, smaller and less specialized, suffered enormous losses but made it through the catastrophe. With the ground cleared of dinosaurs after the dying, mammals evolved rapidly. They increased in body size and took on both plant- and animal-eating lifestyles. Their increased diversity eventually led to the evolution of primates and finally to humans. The mammals inherited the Earth.

It is in that way that planet Earth became the home of intelligent life. You exist, and can read this book, because of an asteroid impact that took place 65 million years ago.

Other Extinctions, Other Impacts?

What about other mass extinctions in Earth history? Can they be connected to cosmic collisions? The answer depends on whom you ask, but the link is not as strong as the impact camp might hope. Of course, the passage of time can erase impact evidence, erode craters, and weather and disturb rock layers. Also, an impactor might cause its own destruction with a shock wave, depositing little iridium and other impact debris.

Of the 24 extinction events that have occurred over the last 540 million years, members of the impact camp claim that 12 are associated with evidence of impacts including iridium, shocked quartz, tektites or craters.

Besides the K-T boundary, the best connection between a large extinction and an impact is at the end of the Eocene, 34 million years ago, when 25 percent of species vanished. Scientists have found a global iridium anomaly and a layer of tektites in sediments from that time. They think that the crater Popigai in Siberia, 100 kilometers in diameter, may be the right age. There may also be a link between extinctions at the end of the Triassic, 212 million years ago, when roughly 60 percent of species were lost, and Manicouagan crater in Quebec, Canada. Shocked quartz and tektites are associated with extinctions at the end of the Pliocene, 2.3 million years ago, and in the middle of the Devonian, 367 million years ago.

There is little evidence to suggest that history's greatest extinction event, when up to 95 percent of species died, was related to an impact. Although it occurred long ago at the Permian-Triassic boundary, 225 million years past, there is no crater and no other impact remains, so far. There is, however, a plausible link between the Permian-Triassic extinction and a flood basalt, the Siberian Traps. The Siberian traps are the largest flood basalt known and they erupted fast, in less than one million years. Their eruption was approximately coincident with the mass extinction. The volcanoes emitted sulfur-rich gases that could have produced global cooling and acid rain.

Of the 12 known flood basalt provinces, at least nine roughly match the time of a mass extinction. The flood basalts with no connection to extinctions, such as the Columbia River Basalts in the Northwestern United States, are relatively small. If flood basalts played a major role in the other mass extinctions, the volcanists argue, why shouldn't they be assigned a dominant role in the K-T extinctions, as well? Perhaps additional research will provide answers to all these questions.

Nemesis—Periodicity in Extinctions

When paleontologist John Sepkoski compiled and reorganized extinction data, he noticed something remarkable. The extinctions of marine organisms seemed to reach a peak every 26 million years in a very regular cycle stretching back 250 million years. There are four major peaks and four smaller ones. Because there were no known geologic cycles with that periodicity, Sepkoski and his colleague, David Raup, issued a challenge to astrophysicists to explain the apparent periodicity. Of the ideas presented, the most popular and plausible was proposed by Richard Muller and colleagues.

Muller hypothsized that our Sun, like most other stars in the Milky Way galaxy, has a companion star. The companion would probably be a faint red dwarf whose eccentric orbit takes it far away from our Sun. Every 26 million years, the star would pass near our solar system, disturbing the comets in the

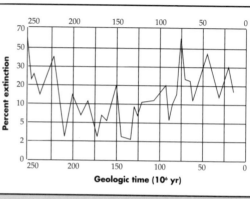

The occurrence of mass extinctions and flood basalts over the past 250 million years.

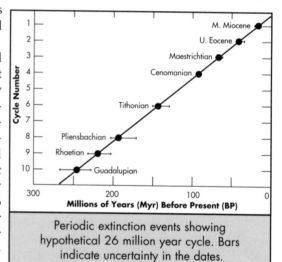

Periodic extinction events showing hypothetical 26 million year cycle. Bars indicate uncertainty in the dates.

Oort cloud and Kuiper belt. Some would spiral in towards the inner planets and one or more could strike Earth, causing a global catastrophe. Muller named our hypothetical companion star Nemesis, the death star.

For years Muller and his small team have searched for Nemesis with no success.

If it exists the star is as distant from the Earth as it ever gets, 13 million years out into its orbit. Despite Nemesis' distance and faintness, the team continues to look—after all, if they were successful, it would be a stunning find. The team has eliminated half of their solar candidates in Northern Hemisphere skies (about 1500 out of 3000), and plans to look in the Southern Hemisphere.

Changing Theories, Changing Philosophies

The transition from theories of global climate change to one of catastrophic meteorite impact to explain the K-T extinctions is a difficult one, and is not yet complete. Scientists have been reluctant to embrace the new idea in part because it represents a basic shift in philosophy. All previous theories involved phenomena found on Earth today—climate, volcanoes, even competition between species. The impact theory posited a catastrophic extraterrestrial cause, one that had more to do with a roll of the dice than with the type of knowable, predictable events scientists have traditionally dealt with.

For decades, evolutionary theory as developed by Charles Darwin held that organisms evolved by the gradual processes of natural selection. The impact theory of extinctions gives chance a much larger role in evolutionary processes. Catastrophism, once associated with the notion that extinctions were caused by the Biblical flood, had been eliminated when Charles Lyell, the father of modern geology, set out the principles of uniformitarianism. The principle of uniformitarianism states that the Earth changes by the same slow and steady processes that are observable today—that the present is the key to the past.

The extinction field of study has been undergoing a scientific "counter-revolution" back to catastrophism. Catastrophism may be difficult to accept not only because it represents a different philosophy, but because it takes something away from our ancestors. In the impactor theory, mammals didn't get ahead because they were better adapted or smarter than reptiles. The Cenozoic reign of the mammals is due to a twist of fate. What is more disturbing is the chance that the history of life on Earth could be radically altered by a catastrophe again.

Twentieth-
Century Hits

Meteorite impacts have defined our planet and its life throughout time. But what cosmic collisions has Earth experienced lately? What effects have collisions had on human life in this century?

Compared with the devastation caused by other recent natural disasters, such as the earthquake in Kobe, Japan, or the eruption of the Mt. Pinatubo volcano in the Philippines, property damage and loss of life from meteorite impacts has been minimal. Nevertheless, meteorite wreckage on Earth is more significant than most people realize—cars, animals, even entire forests have been destroyed by collisions in the twentieth century.

Here are just a few of the important and exciting events of the twentieth century.

Tunguska: Devastation in Siberia

The twentieth century's largest collision took place in a remote region of central Siberia on June 30, 1908. At 7:17 A.M., an enormous fireball blazed through clear early morning skies, witnessed by hundreds of residents of the Podkamennaya Tunguska River Valley. After impact, a column of flame and clouds of thick black smoke billowed up from the site. Hurricane force blasts of hot air knocked people and animals off their feet and broke windows. The native Siberian people, the Evenki, reported that some of their reindeer and dogs died. More than 1,000 square kilometers of forests were flattened.

Evidence of the collision was observed by much of Europe. Explosions were heard over more than one million square kilometers and seismograms recorded earthquake waves more than 800 kilometers to the south. Meteorological stations all over Siberia reported a passing shock wave that was felt more than five hours later in England. It was recorded again 24 hours later in Siberia, after it circled the globe. Light

Downed trees in Tunguska, Siberia, about 5 miles from impact.

reflecting off dust particles blown into the stratosphere illuminated the next several Eurasian nights.

Initial attempts to locate the meteorite failed, since its extreme brightness caused witnesses to underestimate its distance. No official expedition was mounted until thirteen years after the impact, when the Soviet Academy of Sciences sent a party led by Leonid Kulik, curator of the Committee on Meteorites, to the remote and inhospitable region. After four grueling expeditions to the devastated area, during which he dug trenches and drilled holes into the surrounding rock, Kulik found no evidence of a meteorite. The meteorite hunter died in a Nazi concentration camp in 1942 still believing that his bounty was buried in the Siberian wilderness. Fifteen years after his death, the dirt Kulik collected was analyzed by microscope and the first direct evidence of a meteorite impact—tiny spheres of formerly liquid meteoritic dust—was at last observed.

Because there was no meteorite, scientists for decades believed that the impactor was a comet that had vaporized before it struck the Earth. More recently, they have concluded that the object was a stony meteorite, roughly 60 meters in diameter, that exploded and disintegrated at between 5 and 10 kilometers in the atmosphere. The damage was done by the meteorite's shock wave, which struck the ground with the equivalent of a 15- to 20-megaton bomb, about the same amount of energy as was released by the 1980 eruption of Mt. St. Helens in Washington State.

Russian postal stamp commemorating the Sikhote-Alin, Siberia, meteorite fall.

Had the meteorite fallen to Earth only two hours later, it might have hit Moscow. A few hours after that, it would have struck central Europe. Given the enormous distances that celestial objects routinely travel, isn't that uncomfortably close?

A Shower in Siberia

People have observed many meteorites falling, and a few meteorites have later been discovered in the impact craters they created. But in the historic record, only one meteorite has actually been witnessed exploding and creating a large crater field.

On another Siberian morning, this time in February 1947, near the Sikhote-Alin Mountains, hundreds of observers saw a fireball as bright as the Sun fly across the sky. During its brilliant five-second display, it created a kaleidoscope of color, eventually turning bright red. Nearby loggers saw the meteor explode into fragments that descended into the forest with a thunderous crash. Residents felt the shock wave 160 kilometers away. The meteorite shower fell within a field only two kilometers long and one kilometer wide; scientists later calculated that the bolide exploded only about six kilometers above the Earth's surface. In more primitive times, such a sighting might have provoked ritual or religious art—the Soviets struck a commemorative stamp of the fireball.

The impact site little resembled Tunguska. Thousands of meteorite fragments were discovered, the largest weighing more than 1,800 kilograms, some piercing tree trunks and branches. Most of the shards were found in the numerous craters and pits, the largest of which was 26 meters across and 6 meters deep. Interestingly, the largest pieces were not in the biggest craters. The largest meteorites excavated the biggest holes but they shattered on impact, whereas the lower-energy meteorites created smaller craters but remained intact. The total weight of the impactor was estimated at 63,000 kilograms, although only 20,000 kilograms of material has been recovered. The rest vaporized or remains buried in the forest.

The differences in the behavior of the Sikhote-Alin and Tunguska meteorites are primarily the result of their different compositions. Tunguska was probably a stone meteorite while Sikhote-Alin was an iron that did not explode until it was very close to Earth.

Sample of Allende meteorite.

Chevy Malibu after collision with meteorite in Peekskill, New York.

Scientists have been able to calculate the precollision orbit of the Sikhote-Alin meteorite. The object came from the direction of the main asteroid belt, at an aphelion distance of 2.2 astronomical units (AU), but its perihelion was inside Earth's orbit. Could it be that this was an Apollo asteroid inevitably on a collision course with Earth?

Meteorites over Mexico

Those who were still awake at 1:05 A.M. on February 8, 1969, near the village of Pueblito de Allende in the state of Chihuahua, Mexico, were treated to a spectacular sight. A dazzling fireball illuminated the night sky and the landscape below with a pulsating blue-white light. The object broke into two pieces that subsequently exploded into a fireworks display of falling lights. Sonic booms ruptured the night's silence, and thousands of stones fell out of the sky over a region 180 kilometers square.

Afterward, village residents, museum curators and scientists scoured the region for meteorite fragments. In all, more than two metric tons were found. Although no craters were formed and far less meteoritic material was found than at Sikhote-Alin in Siberia, the Allende meteorite has become the most studied. Allende is one of the most important and rare types of meteorites—a carbonaceous chondrite. These black and crumbly rocks have a prominent fusion crust and are brimming with soot-like carbon, possibly the stuff from which life was formed.

Damage to Lives and Property

Meteorites have taken their toll on lives and property in the twentieth century. In the United States alone, there have been more than 20 documented strikes on buildings. Early one morning in 1971, startled residents of a Wethersfield, Connecticut home awoke to find a hole in their roof and a 340-gram stone in their living room. Eleven years later, a 2.5-kilogram stone fell through a roof of another home in the same town! Some scientists think that these events were certainly related; perhaps the meteorites broke off of the same parent body.

Although there is no record of human deaths by stones from space, animals have been killed and at least one person was injured in the past century. Besides the reindeer and dogs who reportedly died in Tunguska, Siberia, a dog was killed by a meteorite fall in Nakhla, Egypt in 1911. One fall day in Sylacauga, Alabama, in 1954, a woman was struck by a meteorite while having an afternoon nap on her sofa. Fortunately for her, the rock was slowed by the roof, the living room ceiling, and a ricochet off her radio before it struck her. Although she lay beneath two quilts, she was badly bruised by the four-kilogram stone. Another stone fell that same afternoon, three kilometers away.

On the evening of October 9, 1992, a bright fireball was spotted flying across much of Kentucky, North Carolina, Maryland and New Jersey. A chondritic meteorite came to rest in a small crater after it plowed through the trunk of a Chevrolet Malibu in Peekskill, New York. When the car's owner discovered the 12-kilogram stone, it was still warm and smelled of sulfur. Because of the uniqueness of the event, the rather ordinary meteorite sold for $69,000 and the car sold for far more than its blue book value.

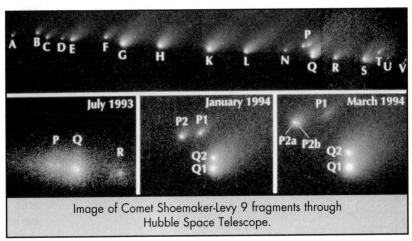

Image of Comet Shoemaker-Levy 9 fragments through Hubble Space Telescope.

Earthlings Witness a Spectacular Collision

Many people thought that a major cosmic collision couldn't happen these days, that the odds were just too small. Luckily, when they were proved wrong, and a major collision did occur, it was with Jupiter. It was the first time scientists predicted a cosmic collision and watched it happen. Although virtually all of Earth's astronomical resources were focused on the giant planet, our most important observer was the Galileo spacecraft, which was only 246 million kilometers away from Jupiter, less than one-third of the distance from Earth. Also, because each of the impacts took place about 10 minutes before that region of Jupiter came into Earth's view, Galileo was the only direct witness.

Comet Shoemaker-Levy 9 was first spotted in March 1993 by Carolyn Shoemaker, who noticed a "squashed comet" in a film of Jupiter taken from the Palomar Observatory telescope in California. The blazing line turned out to be a string of more than 20 comets that had fragmented from a larger body in a close encounter with Jupiter's massive gravitational forces in July 1992. After calculating the comet's orbit, Shoemaker and her colleagues, her husband Eugene, David Levy, and Brian Marsden of the Harvard-Smithsonian Center for Astrophysics, determined that the fragments would strike Jupiter in July 1994.

Scientists debated what to expect from the collisions—a giant impact, with most comet fragments plunging deep into Jupiter's atmosphere, or a big fizzle. Their answer came when the first chunk, Fragment A, traveling at a speed of 200,000 kilometers per hour, struck the planet's atmosphere with the force of 200,000 MT of TNT and sent a plume more than 3,000 kilometers into the stratosphere. Fragment G, thought to be the largest, shot a pock mark of black debris, greater than the diameter of the Earth, into Jupiter's stratosphere. Despite months of study, the size of the original comet remains unknown although each fragment was probably less than one kilometer. Debris from the impacts has entered Jupiter's stratosphere and followed much the same pattern as the ash ejected from Earth's Mt. Pinatubo volcano in 1991. The dark splotches of well-mixed gas and dust have spread longitudinally; some has coalesced and settled out. Scientists predict that once the clouds form a ring at the fragments' impact latitude, they will expand north and south. Within several years the debris clouds will fade from sight; it is

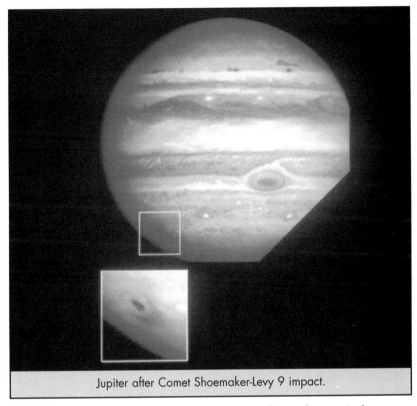

Jupiter after Comet Shoemaker-Levy 9 impact.

already being pulled apart by Jupiter's 300-mile-per-hour winds.

Scientists learned a great deal about cosmic collisions from the Shoemaker-Levy 9 impact. Although comets are just balls of dirt and ice, it is possible for them to strike with a great deal of force. Moreover, it is likely that a large body will fragment in an earlier encounter with its target planet's gravitational forces, reducing the chance of an impact with a giant body but increasing the possibility of impacts occurring in a series—one devastating blow after another. This explains the commonly seen chain of craters on the moon and other planetary bodies.

Perhaps the most important thing scientists, and the rest of the world, learned from the experience of Shoemaker-Levy 9 was that cosmic collisions do occur, and in the time-frame of a human life.

Future
Shock

Scientists were slow to realize the importance of cosmic collisions to life on planet Earth. It was not until 1980, when the Alvarez team showed the world that an extraterrestrial impact may have caused the extinction of two-thirds of the world's species 65 million years ago, that researchers recognized that collisions could shape the course of evolution.

With this realization came questions. If a meteorite could alter the history of life at the end of the Cretaceous, could it do so again? Is there a chance that the reign of humans over the planet could end as abruptly as the reign of the dinosaurs?

And then, in 1994, another sobering event occurred. When Comet Shoemaker-Levy 9 struck Jupiter, it was a startling reminder that the enormous collisions of comets with planets can happen within our lifetimes.

Research into potential collisions with Earth is now a very active field; scientists hold frequent meetings and generate many publications on the subject. Researchers approach this work from three directions. The modelers use the collision records of both the Earth and Moon to determine the likelihood of impacts of various sizes. They calculate the amount of energy released by each collision to determine the consequences of each hypothetical impact. Their frightening scenarios are described in this chapter, which relies heavily on the work of Clark R. Chapman and David R. Morrison.

Another approach to understanding potential colliders is to actually look for them. Astronomers are now searching the skies for potential impactors, believing that we must recognize our enemy before we can conquer it. Finally, both scientists and technologists work to develop methods for dealing with a potential impactor once it has been identified.

What if? Asterion Hermes heads for Manhattan.

Natural Disasters

A cosmic collision is a natural disaster, just like an earthquake, flood or famine. But cosmic collisions are different from other disasters, too. All other natural disasters are limited in magnitude. Geologists believe that the Earth can build up only so much stress before that stress must be released in an earthquake. There are also limits to the strength and size of the storms that accumulate in the atmosphere, even though more people are killed by storms and their resultant flooding than any other natural disaster. Famines may extinguish a large part of a continent's population, but a famine has never reached a global scale.

Cosmic collisions are the only unbounded natural disasters we know of—there is no limit to the size of a potential impactor. While eight natural disasters in the twentieth century have killed between 100,000 and 2 million people, a collision with an object larger than one kilometer in diameter could wipe out billions. Most of the deaths would not result from the impact, but from the aftermath of the collision—starvation due to the collapse of agriculture. Mass starvation would likely precipitate the breakdown of global economic, social and political structures.

Most cosmic collisions would, like other natural disasters, be locally devastating. Locally devastating disasters only injure or kill people who are in the immediate vicinity of the disaster. But a cosmic collision is the one natural disaster that could also be globally devastating. A global catastrophe is one that places everyone on the planet in peril, no matter how far from the impact. By definition, a global catastrophe would cause the death of at least a quarter of the world's population—more than 1.5 billion people today.

The Suspects

To devise a collision scenario, scientists must first determine the rate at which objects of a particular size strike the Earth; that is, their flux. Flux is calculated in two ways. The number of objects striking the top of the Earth's atmosphere is the same as the number hitting the Moon, corrected for the differences in size and gravitational attraction between the two planetary bodies. The flux to Earth's surface must also take into account the effects of the atmosphere on the meteor's survival. Flux can also be estimated by the numbers of meteorite craters of various sizes and ages on the planet's surface.

The extent of the destruction wrought by an impactor depends on several of its qualities including its size, composition, and incoming velocity. Destruction also depends on where it hits—the ocean, uninhabited land, or populated areas. Earth's atmosphere protects us from objects 50 meters or less in diameter, so they do virtually no damage. Small bodies, between 50 meters and one-half kilometer, strike Earth more frequently than larger ones but are only locally damaging. Large meteorites, greater than one kilometer in diameter, are exceedingly rare but have the most devastating effects.

Composition determines the amount of damage a small meteorite can inflict. Small stony meteorites disintegrate before impact, but iron meteorites fragment or hit the surface intact. A larger impactor has so much kinetic energy that its shock wave may cause destruction even if the object itself disintegrates.

Both comets and asteroids are dangerous, but for different reasons. Most of the threat to Earth is posed by the near-Earth asteroids and short-period comets that inhabit the inner solar system, especially those in Earth-crossing orbits. However, long-period comets travel up to three times as fast as asteroids, giving them tremendous kinetic energy. They are likely to explode low in the atmosphere and deliver an enormous shock wave to the surface. Scientists estimate that up to 25 percent of objects that could strike Earth with energies of 100,000 MT are long-period comets.

Long-period comets are also dangerous because they come almost literally out of nowhere. A long-period comet can take more than 2,000 years to complete one orbit, so there may be a comet out there that has never been documented—and that is on a collision course with Earth. With the type of observations currently in use, by the time we detect such a comet it would be only 250 to 500 days from intercepting Earth. One long-period comet probably passes between Earth and Moon each century, and one strikes Earth every few hundred thousand years.

Whether an impactor is an asteroid or a comet, its point of impact is crucial to its potential destructiveness. Most meteorites strike water, since oceans cover 70 percent of Earth's surface. Although small meteorites may fly into the oceans unnoticed, larger impactors could generate tsunamis, imperiling low-lying coastal areas. In fact, the major destruction resulting from an impactor with between 1,000 and 100,000 MT energy (between 100 meters and 1 kilometer in diameter) is likely to be from a tsunami, because the population of the coastal plains is ten times the average land population, and one-third of the world's 50 largest urban areas are found on coastal plains or harbors. Depending on where the impactor hit, a tsunami might destroy up to 100 million people. A meteorite striking land is most likely to hit a sparsely inhabited region, rather than an urban area. Impact location may even determine the magnitude of destruction for a globally devastating collision, as we saw in the increased acid rain that resulted from gases emitted by the limestone at the site of the Chicxulub impact.

Locally Catastrophic Impacts

Most cosmic collisions would cause only local devastation. As we learned earlier, a 60-meter stone struck Tunguska, Siberia, in 1908, and its effects are probably typical for a collision with an object of that type. The meteor disintegrated before it struck the ground, but its shock wave hit with the force of a 15-to 20-MT bomb, destroying more than 1,000 square kilometers of forest. More than 20,000 years ago, another 60-meter object, an iron meteorite, also disintegrated in the atmosphere, but its shock wave created Meteor Crater in Arizona.

One 60-meter object strikes Earth's surface every few centuries, on average. Only once every 100,000 years will one land in a densely populated region. The consequences of a 60-meter object striking an urban area would be extreme. Had Tunguska descended on central Moscow in 1995, instead of on Siberia in 1908, it would have left about nine million people dead. But if an impact of that magnitude occurs in an urban area statistically only once every 100,000 years, that does not rule out the possibility that it will be tomorrow.

What about a larger object, a meteorite 250 meters in diameter with 1,000 MT of energy? If it were to strike the ocean it would propagate a five-meter high tsunami outward from ground zero in a 1000 kilometers radius. Depending on the location of the impact, millions to hundreds of millions of people could be killed. If such a meteorite struck land, it would create a crater five kilometers wide and would destroy 10,000 square kilometers. In one worst-case scenario, a landing in Central Park would wipe out the population of New York City, two-thirds of Long Island, most of Westchester County, a small amount of southern Connecticut, and the most densely inhabited areas of New Jersey. Total casualties would exceed 25 million. Yet few people who were not directly affected by the blast would die and, as horrifying as this scenario is, it is still technically only a local catastrophe. A 1,000 MT meteorite strikes Earth only once every 10,000 years.

Globally Catastrophic Impacts

The transition from a local to a global catastrophe is gradual and not predictable; it depends on several factors. For example, a meteorite impact on rocks that emit acid rain- or greenhouse-generating gases would greatly enhance the number of casualties. Or, the demise of one key species could bring about the collapse of a biological community with globally devastating consequences. Still, modelers calculate that the threshold for a globally catastrophic impact is a meteorite with between 100,000 and 1,000,000 MT energy, about 1.7 kilometers in diameter. The interval between strikes this size or larger is estimated to be about 300,000 years.

On impact, a 1.7 kilometers object would throw enough dust into the stratosphere that the global temperature would drop about 10° C, leading to a world-wide winter and a huge reduction in photosynthesis. Agricultural production might cease for up to a year. Since few nations store such a large food supply, mass starvation would result. The impact would also lead to the destruction of the Earth's protective ozone layer and widespread acid rain.

A larger impact, with energy between one million and 10 million MT, about three to five kilometers in diameter, would completely demolish photosynthesis for months. Gas and dust from the impact re-entering the atmosphere at ballistic speeds would set regional fires. An even larger impactor with around 10 million MT energy would result in global fires. There would be so much dust in the atmosphere that it would be impossible to see. However, it is unlikely that the impacts described would permanently destroy Earth's ecosystems or

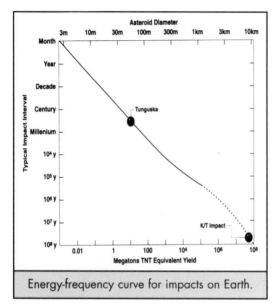

Energy-frequency curve for impacts on Earth.

threaten the survival of the human species.

End of the Cenozoic Collision

The collision at the end of the Cretaceous was more devastating than those described above—it left a different world in its wake. Predictions of what would happen if an equivalent impact were to occur today, that is, a collision that would end the Cenozoic era, have been made by Peter M. Sheehan of the Milwaukee Public Museum and Dale A. Russell of the Canadian Museum of Nature.

Virtually the entire human population would perish, as would the populations of many other species. The impact would trigger continent-wide wildfires, near-total darkness, freezing cold, and intense acid rain. For the survivors of the blast, mass starvation would soon set in. It is likely that all large animals, both wild and domestic, would die of hunger. Just as in the early Tertiary, small animals in detritus-based food chains would have the best chance of survival.

Gradually the environment would become less extreme and some remaining organisms would evolve to replace those that perished. The evolution of life would take unpredictable paths, resulting in very different plants and animals than those that dominate today. We cannot predict who or what those successful organisms might be.

A small number of humans might survive. At first, they would subsist on stored food, which they would unwillingly share with rats and other vermin. Later, human society would need to develop subsistence farming. Although agricultural crop species would survive the holocaust as seeds, farmers would have difficulties growing them since many crops were developed in concert with chemical pesticides and fertilizers. Surviving wild plant species would fare better and some might be domesticated.

It is likely that the K-T impact is the largest collision to have hit Earth in 100 million to one billion years. We hope that it will not be repeated any time soon.

Assessing the Risk

Essentially, this chapter has been about risk assessment, the business of determining how likely an event is to happen and how much damage will be done if it does. Most risks are frequent and have a low consequence, such as the chance of one person falling from a ladder. Some risks are less common but have a high consequence, such as the likelihood of an airplane crash or a hurricane that can kill hundreds at a time. Meteorite impacts are at the extreme end of the rare occurrence with high consequence category.

Risk analysis can be done for meteorite impacts of different sizes. For example, although a Tunguska-like asteroid is destined to reach the Earth once every few centuries, few of the strikes will hit inhabited regions and thousands of years may pass with few deaths. Because of the rarity of strikes on a densely populated area, analysts conclude that on average there are 20 deaths per year by a Tunguska-sized impactor.

The greatest risks are associated with impacts near the global threshold. Although the most likely number of people to die by catastrophic impact in any given year is zero, if one does hit, the consequences are extreme, at least 1.5 billion casualties by definition. If an event this size takes place every 500,000 years, the average global fatality rate is 3,000 people per year.

The chance of an impact with a body the size of the one that ended the Cretaceous is minuscule since one comes around only once every billion years. Even so, with an essentially 100 percent fatality rate, the average number of deaths per year is about 50. Fortunately, the chances of such a disaster are reduced still further, since we know that nothing of that size currently inhabits near-Earth space. At this time, only an impact with a long-period comet could initiate the end of the Cenozoic.

Although one person's annual risk of being killed by a meteorite is only 1 in 1,300,000, over a 65-year life span it's 1 in 20,000—the same chance a resident of the United States has of dying in a commercial airplane accident. Globally, that risk is similar to the risk associated with many natural disasters, but for the average American, the risk of dying by a cosmic collision may be higher than the risk of dying from earthquakes, floods or severe storms.

Since a person stands the same chance of dying from a meteorite impact as in a commercial air accident, some scientists suggest that the global community dedicate roughly the same amount of money now committed to air safety to recognizing and dealing with potential impactors. They suggest a global annual budget for this work of between $10 million and $100 million.

The Unthinkable—Misidentification

One tiny collider, having the energy of the Hiroshima nuclear bomb, one one-hundredth of one MT, strikes the atmosphere each year. A 1-MT hit occurs about once per century. Although these bodies are too tiny to damage the Earth on their own, they can still cause a global disaster. If an unidentified bolide explodes in the atmosphere during a period of high international tension, it might be mistaken for an act of aggression and trigger a nuclear war. The major global powers have satellites that can detect and identify natural high-altitude explosions and so would be unlikely to make this error, but this type of satellite data is not available to nations with emerging nuclear capability.

To avoid this fatal mistake, scientists recommend that the entire globe be alerted to every impact in advance. For this and many other reasons, scientists recommend thorough surveys of the near-Earth skies.

Chances of dying from selected causes (USA).

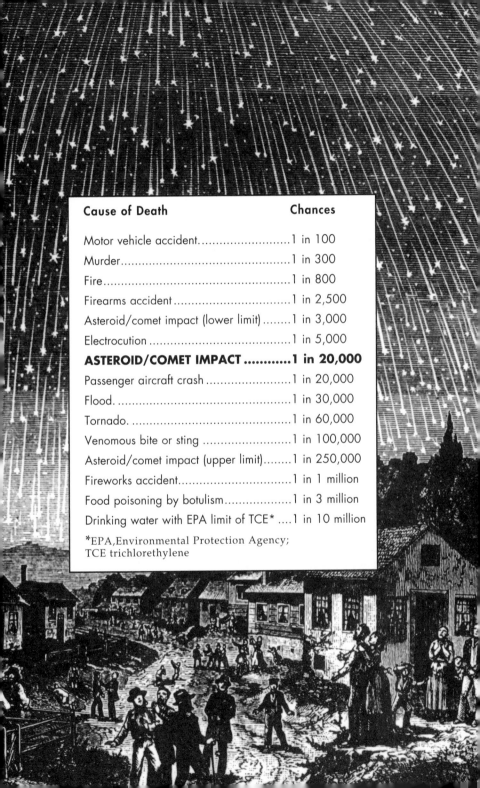

Cause of Death	Chances
Motor vehicle accident	1 in 100
Murder	1 in 300
Fire	1 in 800
Firearms accident	1 in 2,500
Asteroid/comet impact (lower limit)	1 in 3,000
Electrocution	1 in 5,000
ASTEROID/COMET IMPACT	**1 in 20,000**
Passenger aircraft crash	1 in 20,000
Flood	1 in 30,000
Tornado	1 in 60,000
Venomous bite or sting	1 in 100,000
Asteroid/comet impact (upper limit)	1 in 250,000
Fireworks accident	1 in 1 million
Food poisoning by botulism	1 in 3 million
Drinking water with EPA limit of TCE*	1 in 10 million

*EPA,Environmental Protection Agency;
TCE trichlorethylene

Identifying
Flying Objects

Although it is probable that no one will die in a globally catastrophic impact in the next century, the chance that your tombstone will read "killed by a meteorite impact" (provided someone is left to write it) is real. What are scientists and politicians doing to avert the disaster? What should they do?

The first step is to identify the Earth-threatening objects, sometimes called astral assailants. The number of near-Earth objects found increases dramatically as we intensify our search for them. Amazingly, our planet's skies are crowded with approximately 2,000 near-Earth asteroids; some scientists say Earth resides in an asteroid swarm.

Some astronomers, including some amateurs, are dedicated to exploring the skies in order to understand our small-body neighbors. Others calculate the orbits of the newly identified objects to see who our future exterminator might be. As the resources dedicated to the search wax and wane with the political winds and the perceived magnitude of the threat, scientists debate what to do in the future and whether we are doing enough now.

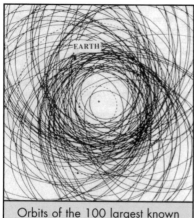

Orbits of the 100 largest known near-Earth asteroids.

Searching and Finding

The search for asteroids and comets has been expanding for two decades, following the realization that an extraterrestrial impact with Earth wiped out the dinosaurs. Advances in technology, as well as the addition of new resources, have increased our ability to find small

Spacewatch Telescope, Kitt Peak, Arizona.

bodies. Old telescopes are being refitted with new detectors to more effectively search the skies for potentially lethal objects.

Small-body optical surveys are performed with two types of detectors —photographic and electronic. In the simplest detector, photographic film is exposed for a period of time and the relative distance and speed of objects is determined by the trail of light left on the film. With a more sophisticated instrument, a charged-couple device (CCD) electronic detector, the night sky is viewed through a telescope and compared with a star map stored in the CCD's computer, allowing it to identify previously unrecognized objects.

For two decades, more asteroids and comets were discovered by astronomers using an old telescope at the Palomar Observatory in southern California than anywhere else. But the Palomar programs have recently been scaled back or halted, because the 59-year-old telescope has only photographic capability and is outdated. It was at Palomar that Carolyn Shoemaker recognized the Shoemaker-Levy 9 comet before it hit Jupiter.

Due to begin in 1996 is a new program at Lowell Observatory, near Flagstaff, Arizona. The Shoemakers will join Edward Bowell in this survey, utilizing an old 24-inch telescope that will be outfitted with a computer drive and state-of-the-art CCD system. The ultimate magnitude of the Lowell endeavor is dependent on funding from the National Aeronautics and Space Administration (NASA).

Complementary to the new program at Lowell is Spacewatch, at Kitt Peak National Observatory in Southern Arizona. While the Lowell telescope will cover a broad region of sky, Spacewatch concentrates on getting a deeper view. Spacewatch, under the direction of Tom Gehrels, has been operating for more than a decade. The program's telescope is frequently upgraded for near-Earth asteroid identification. The Spacewatch enterprise is being expanded; a new larger telescope is to be added to its existing program.

Although a collider is as likely to hit the Southern Hemisphere as the Northern, only one program, at the Anglo-Australian Observatory in Australia, is dedicated to surveying the Southern Hemisphere sky. The program began in 1990 under the direction of Duncan Steel and uses a modern photographic system.

Few observatories in the rest of the world identify small bodies. In Europe, only Czech astronomers search for near-Earth objects. Amateurs contribute a sizable effort to the survey, particularly in Japan and Italy. In fact, more than 20 percent of near-Earth objects (NEOs) have been discovered by amateur astronomers.

After objects are identified, they are reported to Brian Marsden and his team at the Harvard-Smithsonian Center for Astrophysics, who calculate the bodies' orbits and determine what their paths will be for up to the next few hundred years. The results are disseminated to interested scientists around the world. Objects that pass close to Earth are placed on a special list for more detailed observation.

Marsden believes that a major weakness in the current survey system is the inability of one observatory to locate an interesting object spotted by another. Few observatories have suitable equipment for finding a newly discovered near-Earth object and fewer still can halt ongoing projects to do so. Until the ability to relocate objects is achieved, it will be impossible to obtain the observations necessary to deflect an incoming body.

Radar

After an NEO is identified optically, the most powerful ground-based method for determining its physical properties—shape, size, density and rotation—is radar. Radar is also the most accurate technique for calculating the object's trajectory. An orbit based on optical measurements may have an error of up to 100 to 100,000 kilometers (the difference between a hit and a pass several Earth-Moon distances from Earth), but an orbit calculated using radar measurements can determine ground zero to as few as 20 kilometers. Intensive radar surveys could save Earthlings from unnecessary panic a few times every century. Besides being important for orbital calculations, accurate knowledge of physical properties would be crucial to any attempt to defend against an astral assailant.

Currently, two telescopes are equipped with suitable radar capability for NEO observation, the Goldstone telescope in Southern California and the soon-to-be-upgraded Arecibo telescope in Puerto Rico. Because both of these telescopes are oversubscribed, Dr. Steven Ostro, who is responsible for much of the NEO work with radar, recommends bringing at least two new dedicated radar telescopes on-line in the near future. According to Ostro, this project would cost less than one launch of the space shuttle, about $100 to $200 million, yet would produce as much scientific data as a space shuttle mission every week.

Proposed Survey

After the collision of Shoemaker-Levy 9 with Jupiter, NASA formed a committee to develop a proposal for an increased survey of potential impactors. The committee's mission, under the leadership of Eugene Shoemaker, was to propose a relatively inexpensive, cooperative program between NASA and the United States Air Force (USAF), using as much existing equipment with as much international collaboration as possible.

The committee proposed a survey to locate 90 percent of asteroids larger than one kilometer in diameter and 60 percent of all long-period comets over a 10-year period. Astronomers would use existing USAF telescopes, upgraded with high performance CCDs, specifically designed to locate NEOs. However, the USAF currently uses its telescopes for optical tracking of satellites and as of this writing has not yet decided if an NEO search will fit into its program.

This Shoemaker committee proposal came only three years after a previous proposal by another committee of experts, led by David Morrison. Called the Spaceguard survey, this proposal had the more ambitious goal of cataloging 95 percent of all Earth-approaching asteroids and most comets. The Spaceguard committee recommended building six new telescopes and

Meteor over Grand Tetons, Wyoming, August 10, 1972.

suggested a survey period of 25 years. According to Shoemaker, the Spaceguard survey was ideal, but it did not fit into NASA's current budget. Is the scaled-down survey enough? According to Marsden, 99 percent of one-kilometer sized objects within 800,000 kilometers will be detected by the newly proposed survey, but the ability to locate smaller objects will be reduced compared to Spaceguard. He thinks that since small bodies are difficult to recognize in certain parts of their orbits, 10 years is not enough time to locate all potential assailants. Also, the USAF telescopes are too small to observe all the important bodies, especially long-period comets, which approach rapidly from a great distance.

Shoemaker believes that, with the anticipated improvements in CCD technology, the proposed program will meet its goals at a reasonable cost. Although the ability to identify some extraterrestrial threats is reduced, the program is realistic given current budget constraints.

Of course, even the new proposal is just a proposal; its fate depends largely on the amount of money NASA is willing and able to spend on a near-Earth object survey.

Close Encounters

Astronomers have never identified anything in space before it hit the Earth or even entered our atmosphere. Comet Shoemaker-Levy 9 was the only object ever recognized before a cosmic collision. But scientists, with their improved equipment, continue to look for potential colliders and, in the process, are finding smaller, fainter and closer objects all the time.

In May 1994, Tom Gehrels turned the Spacewatch telescope to the night sky and discovered the closest object to Earth ever observed. The six-meter body was spotted moving away from us after passing 150,000 kilometers from our planet. The week Gehrels received his copy of the 1994 *Guinness Book of World Records* acknowledging his discovery, one of his Spacewatch partners, Jim Scotti,

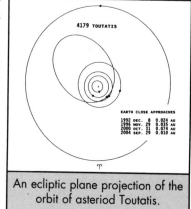

An ecliptic plane projection of the orbit of asteriod Toutatis.

located a nine-meter object only 105,000 kilometers away. Undoubtedly this record will fall many times.

Asteroid 4179 Toutatis is the most (potentially) dangerous object currently recognized. Its chaotic orbit brings the two-to three-kilometer object very close to Earth once every four years. In December 1992, Toutatis came within about 3.6 million kilometers of our planet. It will approach about 1.5 million kilometers in September 2004. Calculations of the asteroid's orbit out to 2200 A.D. indicate that it is not likely to strike Earth, although its orbit is so chaotic that it is difficult to be certain more than 100 years in advance. Astronomer Keith Noll predicts that over the next 10 million years Toutatis will collide with Earth or be knocked out of the solar system by the gravitational field of a large planet.

The consensus among astronomers is that, in the next 200 years, it's unlikely that an object a few kilometers or larger will strike Earth since a candidate has not been found. But only about 5 percent of one-kilometer objects have been identified so far, so their threat is not well understood. And the majority of smaller objects, those that could cause local damage, will take decades to identify.

Of course, long-period comets are a different problem. The more sensitive the optical survey scientists engage in, the farther away they can detect a long-period comet and the longer warning period we would have before a potential collision. As Eugene Shoemaker says, "the only thing we can do about long-period comets is maintain eternal vigilance."

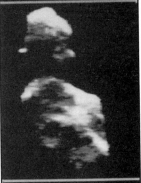

Goldstone radar images of asteroid 4179 Toutatis.

Defensive
Maneuvers

Suppose for a minute that all the astronomical resources of our planet indicate that a massive object is on a collision course with Earth. What do we do? What can we do?

Our response, of course, depends on the object— what it is, how big it is, how fast it's traveling, how long we've known about it, and how soon it will be here. Our response will also depend on the preparations we've made for a defense system.

The worst case is that a large, fast object hits with little warning. If tonight, a long-period comet were discovered headed straight for Earth, impact would be likely two months to two years away. There's not much we could do on such short notice. Radar would determine the point of impact within 50 kilometers, and an evacuation plan could be formulated for people at ground zero and in low-lying coastal areas, if a tsunami were predicted. Any remaining residents, the local environment and all nonmovable property would be destroyed. If ground zero were a small country, say Israel, the evacuation of a large segment of the population might be politically challenging.

This scenario is probable now, when we have little ability to defend ourselves against an impactor, and may be likely for a long time into the future. Opinions vary on how much preparation we should make and on how much risk we as a society are willing to accept. Most astronomers think there is little need to prepare for an impact until a definite threat has been identified, and that we should accept the small risk that an impactor will emerge from nowhere with little warning. Other scientists and technologists think that we need to be prepared for any cosmic collision, even with a rapid response to a newly identified speeding comet.

> "End of the World"
> by Thomas Voter.

Preemptive Visits

The most important element of defense against an astral assailant is knowledge. Scientists need to locate potential impactors and intimately understand their movements in the vicinity of Earth. In addition to Earth-based surveys, many scientists favor expeditions to near-Earth objects (NEOs) in order to better understand the bodies before we have to deflect one, and to better understand our solar system.

Some scientists suggest we make visits to asteroids for friendly purposes. They argue that long before we locate an astral assailant we should learn to use asteroids for their resources and protection. For example, space travellers could mine distant asteroids for metals and water and could also use them to provide protection from solar radiation and cosmic rays for space ships or colonies.

In addition, rendezvous missions to NEOs would gather the information about mass, density, internal structure, etc., necessary for accurate deflection of a body moving towards Earth. Knowing the range of these properties from visits to nearby objects might help scientists design an effective and rapid defense response, should one be necessary, to an unexplored body. Methods for intercepting, analyzing and landing on an asteroid could also be tested during the missions.

Ideally, an astral assailant would be identified in time for scientists to mount a rendezvous expedition. By thoroughly understanding the object's composition and density, they could determine the magnitude of its threat; for example, whether it would explode in the atmosphere or strike the ground. If defensive maneuvers were required, the information could be used to determine the most effective method.

February 16, 1996, is the launch date for the first scheduled rendezvous with an asteroid (the visits to Gaspara and Ida discussed in Chapter 4 were fly-bys). The target is 433 Eros, the second largest near-Earth asteroid. After its arrival at the target site in January 1999, the spacecraft will spend one year orbiting within 50 kilometers of the irregularly shaped body (36 by 15 by 13 kilometers). The craft will perform experiments to determine the asteroid's composition, internal structure, and geologic history. Although not considered a major future threat, 433 Eros ventured to within 22 million kilometers of Earth in 1975.

Defense Strategies

The time between the discovery of an impactor and its collision date determines the types of defenses that may be engaged. If action could be taken at least several orbital periods in advance of an impact, an attractive defense would be to deflect the object from its orbit. Only a small impulse would be required to push, or deflect, a small-to-medium body from its orbit many decades prior to its encounter with Earth. For these situations, conventional chemical weapons could be used.

For bodies too large or too close to impact to be easily deflected, the choice between more aggressive means of deflection and destruction, also called fragmentation, would be necessary. Assailants larger than 100 meters in diameter would probably require nuclear explosives, but the launch and delivery of a nuclear warhead to the target would need to be carefully designed. Even if it were possible to develop a defense system in time to deflect an incoming object, evacuation might be chosen over a nuclear explosion if the object was expected to produce only local devastation.

The estimate is that the development of a defense system against an astral assailant would take 10 years—a rapid enough response to neutralize most asteroids and short-period comets. Many existing technologies and some existing equipment could be used. For example, if a warhead is needed to deflect the object, a large Russian launcher could carry an interceptor loaded with an American nuclear warhead into space. The global scientific community could develop the technology necessary for navigating and targeting from current asteroid and satellite tracking systems, the horror of a threat from outer space neutralizing any political obstacles to collaboration. With enough time before impact, new technologies with new equipment specifically designed for asteroid and comet interception could also be developed.

Deflection

Even a small change in orbital velocity could divert an asteroid or comet from a collision with Earth if it were done well ahead of interception. Changing an object's velocity requires changing its kinetic energy, which can be altered by a change in its mass. An assailant's mass could be decreased by sending a device to its surface or by working remotely.

The only proposed ground-based deflection system is an enormous laser that would direct massive amounts of energy to one side of the target. The laser beam would increase the temperature of one side of the body's surface layer, causing it to crack and ultimately to split off, decreasing the object's mass and forcing it into a different orbit. This technique would require a great deal of lead time since the technology is undeveloped.

A few decades would be needed to develop and deploy a mass-driver system. A mass driver, placed on the asteroid, would mine material from one side of the body and toss it into space over a number of years or decades. Continued shooting material of the body would ultimately change its trajectory. Delivering a mass driver to an asteroid's surface and using it are formidable engineering tasks and are beyond our current capabilities.

There are two types of explosions that could deflect an approaching assailant. Either might use conventional or nuclear warheads, although nuclear bombs contain more than a million times the energy of a conventional weapon of the same mass.

One type of explosion, called a standoff explosion, is similar to the laser-beam method discussed above. Energy from a bomb would increase the temperature of one side of the outer layer of the target, causing it to break off, altering the asteroid's orbit. The most efficient type of standoff explosion is a nuclear bomb detonated a few hundred meters above the target. Scientists calculate that up to 10 kilotons of explosives would be needed to deflect a one kilometer body and 10 megatons would be needed to deflect a 10-kilometer-wide body. This technique is thought to be the most energy-efficient for deflecting a potential impactor.

An impactor might also be nudged from its orbit by a surface explosion that results in cratering. This technique would only work for small objects that pose no real threat to Earth anyway because surface explosions on larger objects might cause them to fragment into large pieces, each of which would enter an unpredictable orbit and pose its own threat to Earth.

Fragmentation

It would seem that the most obvious way to avoid a cosmic collision would be to destroy the impactor. However, the problems arising from the generation of large unpredictable fragments in an explosion are almost insurmountable. To destroy an impactor, an explosion must be so large that each resulting fragment is smaller than 10 meters, assuring that it will burn up in the Earth's atmosphere. The best way to achieve optimum fragmentation is to bury the explosive deep in the impactor. This technique requires a device to bury the charge (currently undeveloped), as well as a huge amount of energy to blow the body up. A 1-MT warhead could destroy a 750-meter sphere, and a 1,000 MT bomb could destroy a seven-kilometer body. Besides being dangerous, fragmentation is much less energy-efficient than deflection.

One of the more unusual suggestions for destroying an asteroid was termed Brilliant Mountains, the big brother to the controversial Star Wars proposal, Brilliant Pebbles. In this method, small asteroids would be nudged into exactly the right Earth orbit to attack a bigger asteroid. The proposal has not received serious consideration.

All of these systems are more suitable for a defense against asteroids than against long-period comets. Comets come from far away and are very fast. Probably the only way to eliminate the threat from a long-period comet would be to disintegrate it, a task that would take a prohibitively large amount of energy.

How Much to Prepare

Most astronomers think that it is premature to invest a large effort in developing a defense system against an astral assailant until a hazard is defined. Instead, they suggest a modest attempt to consider technologies that might be employed, but discourage actual development of those technologies at this time. Since most of the hazards from asteroids can be predicted more than 20 years in advance, they say that the risk of waiting to develop defensive technology is not great. If an assailant with an estimated time of arrival of 20 years were discovered, astronomers advocate making measurements, exploring the body with a probe and only then developing a system to deflect it.

The End?

The history of the Earth is a history of cosmic collisions. From the beginning of time, these solar system processes have affected much about our planet—its formation, the birth of its Moon, and the evolution of its life. Most sobering is the recognition that life was fundamentally altered by an impact at the end of the Cretaceous. We know that sometime in the future, a destructive collision with Earth will occur again.

Yet, we do not live in the thousand-or million-year time frames that are the context for large impacts. Although we understand that there is a statistical chance of a significant collision happening tomorrow, we should also know that it is exceedingly small. We have seen no major crater-forming impact in recorded history and we know of no documented case of a human death by a meteorite.

Many other urgent hazards face human society: overpopulation, global warming, ozone depletion, deforestation, malnutrition, disease, pollution, nuclear proliferation, ethnic conflicts, and racial strife. The remote threat of a globally devastating cosmic collision serves to remind us of the more immediate threats to our planet. We recognize that the global warming produced by the release of carbon dioxide into the atmosphere by the Chicxulub impact is similar to the climate change induced by the CO_2 emissions of industrial society. The stripping of the ozone layer by the plume of dust and gas that likely followed the collision at the K-T boundary is reminiscent of the ozone hole that grows during Antarctic winters. The model of a global winter resulting from dust kicked into the atmosphere by the K-T impact brings to mind models of a winter caused by nuclear war. These environmental disasters and threats are real, without the threat of a major meteorite impact.

The risk of death by a meteorite impact is statistically measurable for each of us, but in a world with limited financial resources, how much should be put into identifying and preparing for a potential collider? Many astronomers suggest that resources should match the degree of risk. The chance that any individual will die of cancer is far greater than the chance that she will die from a meteorite impact; more money should be put into cancer research than anti-meteorite defenses. The risk of death in a commercial airplane accident matches the meteorite risk; perhaps funding to address these risks should be equal.

Certainly it would be prudent for astronomers to continue surveying the skies for potential assailants. Visits to near-Earth asteroids would provide data on the makeup and orbital paths of our neighbors that would be useful should a defense against an extraterrestrial threat become necessary. As an added bonus, many fundamental questions about the solar system could be answered relatively inexpensively through missions to these poorly known objects.

Long-period comets pose a different problem, since they can be recognized no sooner than two years before impact. Any system capable of neutralizing a comet would have to be quite advanced; it would take well over two years to develop, and might not be current by the time it was needed. How can governments sustain funding against such an unpredictable threat when more immediate problems continually arise? Should we move now to produce a defense system to deal with this threat even though it would likely involve enormous nuclear warheads?

Most astronomers say no. At some level, society has to live with risks. We take chances daily when we drive our cars, walk down the street, or even eat a cheeseburger. Many Californians live with the ominous threat of earthquakes, yet few people leave the state to avoid them. Still, it is distressing to know that the only course of action available today for an impending long-period comet is evacuation of the target area (which would save no one if the impactor were large enough to cause global devastation).

What if a collision were inevitable? Might something good come from it? Astronomer Keith Noll admits being disappointed when new calculations revealed that Comet Swift-Tuttle would not strike Earth, after the 1992 prediction that it would precipitate "the end of the world" in 2126. He wonders, "would dealing with an enormous extraterrestrial threat cause us to put a hold on racial tensions, ethnic strife, and political problems? Would we quit blowing each other up and work together to save our collective hides?"

We may only know the answer if Arthur C. Clark's vision of a global threat of annihilation is more than just science fiction.

overleaf:
Meteorite
streaks to
Earth.

P H O T O   C R E D I T S

p.2—William K. Hartmann

p.3—Institute of Meteoritics, University of New Mexico

p.4—Picture Collection, The Branch Libraries, The New York Public Library

CHAPTER 1
p.11—William K. Hartmann
pp.12, 13—Negative #331534, Photo: Barnum Brown Collection, Courtesy of Department of Library Services, American Museum of Natural History

CHAPTER 2
p.15—Picture Collection, The Branch Libraries, The New York Public Library p.16—Diagram courtesy of Sky Publishing Corp. p.17 (A & B)—"The Earth's Elements", Robert P. Kirshner. © Oct. 1994 by Scientific American Inc. All Rights Reserved. Photo: J. Tesler, Arizona State University/NASA p.19—Picture Collection, The Branch Libraries, The New York Public Library pp.20, 21—NASA p.23—O. Richard Norton, Rocks From Space, Mountain Press Publishing Co.; Illustration by Dorothy S. Norton

CHAPTER 3
p.25—July 1994 Scientific American Cover. © July 1994 by Scientific American Inc. All rights reserved. Artist: Alfred T. Kamajian p.26—William K. Hartmann pp.27, 29—NASA pp.32, 33—Photo by J. D. Griggs/ U. S. Geological Survey

CHAPTER 4
p.34—R. Kempton (New England Meteoritical Services), Space Telescope Science Institute p.35—Negative #313171, Photo: Prague Observatory, Courtesy of Department of Library Services, American Museum of Natural History pp. 36-37—JPL/NASA p.40—NASA p.42—O. Richard Norton, Rocks From Space, Mountain Press Publishing Co. Illustration by Dorothy S. Norton p.43—JPL/NASA, Image courtesy of Dr. Steven Ostro

CHAPTER 5
p.47—Shin Photos p.48—Picture Collection, The Branch Libraries, The New York Public Library p. 49 (top)—Negative #323886, Photo courtesy of Department of Library Services, American Museum of Natural History p. 49 (bottom)—Negative #601574 13/13A, Photo by Cone/Goldberg, Courtesy of Department of Library Services, American Museum of Natural History p.50 (top)—Photo courtesy of Ursula Marvin, Smithsonian Astrophysical Observatory p.50 (bottom)—Bruce Bohor/USGS p.51—USGS pp.52, 53—Institute of Meteoritics, University of New Mexico

CHAPTER 6
pp.57-59—Geological Survey of Canada p.60—Geological Survey of Canada, Photo: R. Grieve p.62—"Impact Cratering on the Earth", Richard A. F. Grieve. © April 1990 by Scientific American Inc. All rights reserved. Illustration by Ian Worpole p.63 (top)—Dr. L. Crossey, University of New Mexico p.63 (bottom)—Geological Survey of Canada p.65—Negative #322441, by Paula Hutchinson, Museum Art Dept., Photographed by L. Boltin, Courtesy of Department of Library Services, American Museum of Natural History

PHOTO CREDITS

CHAPTER 7

p.67—*Negative #323617, Courtesy of Department of Library Services, American Museum of Natural History* p.68—*Philip Sze/Dept. of Biology, Georgetown University* p.69—*Negative #45558, Photo by E. O. Hovey, Courtesy of Department of Library Services, American Museum of Natural History* p.70—*Walter Alvarez, University of California, Berkeley/Department of Geology and Geophysics* p.71—*"An Extraterrestrial Impact", Walter Alvarez.* © *Oct. 1990 by* Scientific American Inc. *All rights reserved. Diagram by George Retsick* p.72 (top)—*Negative #2A7681, Photo by Rota, Courtesy of Department of Library Services, American Museum of Natural History* p.72 (bottom)—*An Extraterrestrial Impact, Walter Alvarez.* © *Oct. 1990 by* Scientific American Inc. *All rights reserved. Diagram by George Retsick* p.73—*Geological Survey of Canada* pp.74, 75—*J. Mahoney, University of Hawaii* p.76—*Ross Griffiths, The Australian National University* p.77—*The New York Times/Graphic by John Papasian*

CHAPTER 8

p.81—*Negative #322113, Photo by Boltin, Courtesy of Department of Library Services, American Museum of Natural History* pp.82, 83—*Negative #2A7603, Photo by Logan, Courtesy of Department of Library Services, American Museum of Natural History* p.86—*Reprinted from* The Nemesis Affair: A Story of the Death of Dinosaurs and the Ways of Science, *by David M. Raup, with permission of W. W. Norton & Co., Inc. Copyright* © *1986 by David M. Raup*

CHAPTER 9

p.89—*Negative #126035, Photo courtesy of Department of Library Services, American Museum of Natural History* p.90—*O. Richard Norton,* Rocks From Space, *Mountain Press Publishing Co., Photo* © *O. Richard Norton* p.91—*Institute of Meteoritics, University of New Mexico* p.92—*Photograph by Annika K. Johansson* pp.93, 95—*Photos courtesy of Sky Publishing Corp.*

CHAPTER 10

p.97—*Negative #336042, Courtesy of Department of Library Services, American Museum of Natural History* p.102—*Courtesy of Dr. Morrison, NASA/AMES* p.105—*Negative #119137, Courtesy of Department of Library Services, American Museum of Natural History*

CHAPTER 11

p.106—*Courtesy of R. P. Binzel/MIT* p.107—*Negative #124651, Courtesy of Department of Library Services, American Museum of Natural History* p.108—*James V. Scotti, University of Arizona* pp.110, 111—*James M. Baker* p.112—*Reprinted from* Hazards Due To Comets and Asteroids, *T. Gehrels editor, The University of Arizona Press, Copyright* © *1994* p.113—*JPL/NASA, image courtesy of Dr. Steven Ostro*

CHAPTER 12

p.115—*Negative #315901, by Thomas Voter, Photo taken by Thane Bierwert, Courtesy of Department of Library Services, American Museum of Natural History*

CHAPTER 13

pp 122, 123—*William K. Hartmann*

I N D E X

243 Ida, 39, 116
433 Eros, 116
951 Gaspara, 39, 116
1580 Betulia, 44
1627 Ivar, 44
1915 Quetzalcoatl, 45
1981 Midas, 45
4179 Toutatis, 113
4769 Castalia, 45
achondrites, 53
acid rain, 83
Alba Patera, 79
Allende, 92
Alvarez, Luis, 71
Alvarez, Walter, 71
American Museum of Natural History, 50
Amors, 44
Anglo-Australian Observatory, 109
angular momentum, 18
Anhingito, 50
Antarctica, 52
Apollo missions, 56
Apollos, 44
Arecibo telescope, 110
Asaro, Frank, 72
asteroid belt, 21, 36
asteroids, 21, 34-45
asteroids, composition, 38
astroblemes, 56
Atens, 44
AU (astronomical unit), 36
Barringer, Daniel M., 64
Barringer's Crater, 64
Big Bang Theory, 14
blast magnitude, 62
bolides, 49
Bowell, Edward, 108
Brilliant Mountains, 119
Brilliant Pebbles, 119
Canadian Geological Survey, 58
Canadian Museum of Nature, 102
Cape York meteorites, 50
carbon dioxide, 83
carbonaceous chondrites, 54, 92
Cenozoic Era, 69, 102
Ceres, 36
Chapman, Clark R., 98
Chicxulub, 61, 75
chemical weapons, 117
Chevrolet Malibu, 93
chondrites, 53, 73, 92
Clark, Arthur C, 10, 121

Comet Shoemaker-Levy 9, 6, 11, 94, 108
Comet Swift-Tuttle, 8, 121
comets, 21, 34-45
Committee on Meteorites, 90
complex craters, 62
cratons, 60
Cretaceous-Tertiary boundary, 68
Dactyl, 39
Darwin, Charles, 87
Deccan Traps, 78
defense strategies, 117
deflection, 117
detritus, 82
Devonian Period, 85
differentiation, 22
dinosaurs, 7, 10, 66-79, 80
dirty snowballs, 41
Earth, atmosphere, 31
Earth, structure of, 23
Eocene, 85
evolution of life, 7
evolutionary theory, 87
extinct comets, 45
falls vs. finds, 51
fireballs, see meteorites
flood basalts, 76, 85
flux, 98
fragmentation, 117, 119
galaxies, 16
Galileo spacecraft, 39, 94
Gehrels, Tom, 108, 112
geological time scale, 66
Giotto, 45
global catastrophe, 98
globally catastrophic impacts, 101
Goldstone telescope, 110
gravitational potential energy, 18, 22
Grieve, Richard A.F., 58
ground-based deflection, 118
Gubbio, 71
Guiness Book of World Records, 112
Halley's comet, 31, 42, 73
Halley's comet, 31, 42, 73
The Hammer of God, 10
Harvard-Smithsonian Center for Astrophysics, 109
helium, 16
Hellas Plenitia, 79
Hoba, 10, 50
homogenization, 22
hydrogen, 16
igneous asteroids, 38

impact basalts, 31
impact craters, 56
impactor theory, 26, 87
Jefferson, Thomas, 46
Jupiter, 6, 94
Kali, 10
kinetic energy, 18, 22, 61, 63, 117
Kitt Peak National Observatory, 108
Kuiper belt, 41, 86
Kuiper, Gerard, 41
Kulik, Leonid, 90
Levy, David H., 94
locally catastrophic impacts, 100
long-period comets, 41, 99
Lowell Observatory, 108
lunar crater, 28
Lyell, Charles, 87
mammals, 84
Manicouagan, 61, 85
Mars, 53, 79
Marsden, Brian, 8, 94, 109
mass driver, 118
mass extinctions, 7, 66
Mesozoic Era, 68
metamorphic asteroids, 38
meteor, 46
Meteor Crater, 10, 64
meteor showers, 50
meteorites, 34, 46-55
meteoroid, 46
Michel, Helen V., 72
micrometeorites, 53
Milky Way galaxy, 18
Milwaukee Public Museum, 102
Montagnos, 61
Moon, 10, 24
Morrison, David R., 96, 110
Muller, Richard, 86
National Aeronautics and Space Administration (NASA), 108, 110
natural disasters, 98
natural selection, 87
near-Earth asteroids, 38, 99
near-Earth objects (NEOs), 44
Nemesis, 86
Noll, Keith, 113, 121
oceans, 31
Oort cloud, 40, 86
Oort, Jan, 40
Ostro, Steven, 110

Palomar Observatory, 94, 108
Peekskill, 93
Permian-Triassic boundary, 85
Perry, Robert E., 50
phytoplankton, 69
Piazza, Guiseppe, 36
planetesimals, 19
planets, 19
plate tectonics, 28, 57
Pliocene Period, 85
Popigai, 61, 85
primitive asteroids, 38
Quaternary Period, 68
radar, 114
radioactive clocks, 55
radioactive decay, 54
Raup, David, 86
Russell, Dale A., 102
Scotti, Jim, 112

Sepkowski, John, 85
shatter cones, 63
Sheenan, Peter M., 102
shock, 63
shocked quartz, 63, 74
Shoemaker, Carolyn, 94, 108
Shoemaker, Eugene, 64, 94
shooting stars, see meteorites
short-period comets, 42, 99
Siberian traps, 85
Sikhote-Alin, 61, 91
simple craters, 61
The Simpsons, 10
small bodies, 34
Soviet Academy of Sciences, 90
Spaceguard survey, 110
Spacewatch, 108
standoff explosion, 118
Star Trek, 10

Steel, Duncan, 109
sublimation, 40
Sudbury Structure, 10, 60
Sun, 14, 86
Sylacauga, 93
tektites, 63, 75
Tertiary Period, 68
The Hammer of God, 10
Titus, John D., 36
Triassic Period, 85
tsunamis, 75, 100, 114
Tunguska, 88-90
uniformitarianism, 87
Viking spacecraft, 53
volatiles, 19, 31
volcanoes, 76-78
Vredefort Rinf, 60
Wethersfield, 92
Willemette meteorite, 50

F U R T H E R R E A D I N G

Alvarez, Luis, Walter Alvarez, Frank Asaro, and Helen V. Michel, "Extraterrestrial Cause for the Cretaceous-Tertiary Extinction," *Science*, June 1980

Binzel, Richard P., M. Antonietta Barucci, and Marcello Fulchignoni, "The Origin of Asteroids," *Scientific American*, October 1991

Burnham, Robert, "Arizona's Meteor Crater," *Earth*, January 1991

Canavan, Gregory H. and Johndale Solomen,"Interception of Near-Earth Objects," *Mercury*, May-June 1992

Clarke, Arthur C., *The Hammer of God*, Bantam Books, 1993, 240pp

Gehrels, Tom, ed., *Hazards Due to Comets and Asteroids*, University of Arizona Press, 1994, 1300pp

Glen, William, "What Killed the Dinosaurs?," *American Scientist*, July-August 1990

Hartmann, William K and Ron Miller, *The History of the Earth*, Workman Publishing, New York, 1991, 260pp

Morrison, David and Clark R. Chapman, "Target Earth: It Will Happen," *Sky & Telescope*, March 1990

Morrison, David "The Spaceguard Survey: Protecting the Earth from Cosmic Impacts," *Mercury*, May-June 1992

Norton, O. Richard, *Rocks from Space: Meteorites and Meteorite Hunters*, Mountain Press Publishing Co., 1994, 449pp

Raup, David M., *The Nemesis Affair: A Story of the Death of Dinosaurs and the Ways of Science*, W.W. Norton & Co., 1986, 220pp

Ward, Peter, *The End of Evolution: On Mass Extinctions and the Preservation of Biodiversity*, Bantam Books, 1994, 302pp

Wilhelms, Don E., *To a Rocky Moon: A Geologist's History of Lunar Exploration*, University of Arizona Press, 1993, 477pp

A C K N O W L E D G M E N T S

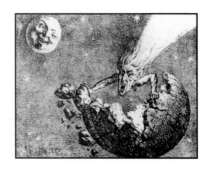

Special Thanks to:

*I would like to thank, above all, the scientists who have dedicated their lives to the
study of Earth, space and mass extinctions. It is they who discover and share with us
the story of our world. Conversations with some of them—Eugene Shoemaker, Tom
Gehrels, Heidi Hammel, Brian Marsden, Keith Noll and Steven Ostro—contributed
to the quality of this work. I also received counsel from Walter Alvarez, Edward
Anders, Vincent Courtillot, Richard Muller and Norman Sleep. Many thanks to the
staff of Robert Ubell Associates, especially Mark Meade who introduced me to the project, and to Luis Gonzalez who saw me through it. Thanks also to Bob Ubell who was
always encouraging. I am also grateful to JC Suares and Gillian Theis who worked so
hard to provide the right art and design. Several scientist friends were invaluable in
providing or tracking down important materials. Special thanks to Debra Colonder
and John Longhi of the Lamont-Doherty Earth Observatory of Columbia University,
and Rosamund Kinzler, Albert Leger and Edward Mattez of the American Museum of
Natural History. Significant contributions to editing this work came from David
Sobel of Owl Books, Corey S. Powell of* Scientific American, *and from Walter
Crokett of* Worcester Magazine *who provided advice and encouragement. In addition
to all the things spouses are usually thanked for in books—love, support, patience—
Miles Orchinik must also be acknowledged for helping to make sense of a very early
version of the manuscript. Without him, my job would have been much more difficult.
This book is dedicated to the memory of my father, James Joseph Desonie.*

— DANA DESONIE